The F-EIR Conference 2021 on
Environment Concerns
and its Remediation

Volume of Abstracts

The F-EIR Conference 2021 on
Environment Concerns and its Remediation

Volume of Abstracts

Edited by

Deepankar Kumar Ashish

Organized by

Federation of Environmental Issues and its Remediation (F-EIR)
In association with Maharaja Agrasen University

Supported by

Science of the Total Environment, an Elsevier journal
Environmental Science and Pollution Research, a European Chemical Society's Journal
Sustainability, a MDPI journal
Macromolecular Symposia, a Wiley Journal
Materials Today Proceedings, an Elsevier journal
Lecture Notes in Civil Engineering, a Springer Bookseries
Springer

ALLIED PUBLISHERS PVT. LTD.

New Delhi • Mumbai • Kolkata • Chennai • Bangalore • Hyderabad

Editor
Deepankar Kumar Ashish
Department of Civil Engineering
Maharaja Agrasen University
Baddi, Himachal Pradesh, India

ISBN: 978-93-90951-26-0

Allied Publishers Private Limited
D-5, Sector-2, Noida 201 301
Ph. Nos.: 0120-4320295/2542557/4352866
E-mail: delhi.books@alliedpublishers.com
www.alliedpublishers.com

Preface

The F-EIR Conference 2021 – Environment Concerns and its Remediation was held on 18–22 October in Chandigarh, India. The event was aimed to bring research professionals from multi-disciplinary fields to cross established sub-disciplinary divides, encouraging the exchange of ideas between scientists, engineering professionals, architects, environmental scientists, academicians, economists, and students. The conference focussed on the most interesting and relevant critical thinking on environmental issues with a wide array of quality technical presentations.

Over 400 abstracts and 300 full papers were received by the Organizing Committees, and about 140 paper were finally accepted for presentation in 27 sessions of F-EIR Conference 2021. These papers were presented by world renowned experts from 30 countries during the event.

The abstracts of papers presented are published in Volume of Abstracts, and the online proceedings contains all the accepted papers including 10 keynote lectures. Some selected papers will appear in the Science of the Total Environment, an Elsevier journal having Impact Factor 7.963, Environmental Science and Pollution Research a European Chemical Society's journal published by Springer journal having Impact Factor 4.223, Sustainability a MDPI journal having impact factor 3.251, Macromolecular Symposia a Wiley journal, Materials Today Proceedings an Elsevier journal, Lecture Notes in Civil Engineering a Springer bookseries and book volume in Springer.

As the chairman of the conference, I would like to extend my appreciation to all the participated authors of their valuable contributions to the conference which made F-EIR Conference 2021 a most successful one. Special thanks are due to the keynote speakers, and scientific and advisory committee members for their hard work and time spent to make this conference more interesting.

Chandigarh, India
October 18, 2021

Deepankar Kumar Ashish
Chair, F-EIR Conference 2021

Organization of Conference–
The F-EIR Conference 2021

Conference Chair
Deepankar Kumar Ashish

Secretary General
Bobby

Advisory Committee

Dionysios (Dion) D. Dionysiou, United States
Abdul Rashid Dar, India
Prashant Kumar, United Kingdom
Lucian Lucia, United States
Surinder K. Mehta, India
Lidia Morawska, Australia
Hai Nguyen Tran, Vietnam

Scientific Committee

Dimitrios G. Aggelis, *Belgium*
Carmen Andrade, Spain
Deepankar Kumar Ashish, India (Chair)
Dipendu Bhunia, India
Paulo Barreto Cachim, Portugal
Antonio Caggiano, Germany
Aires Camões, Portugal
Fernando Chiñas-Castillo, Mexico
Luigi Coppola, Italy
Marco Corradi, United Kingdom
Martin Cyr, France
Sławomir Czarnecki, Poland
Pedro Raposeiro da Silva, Portugal
Jian-Guo Dai, Hong Kong
Joaquim António Oliveira de Barros, Portugal
Frank Dehn, Germany
Enrico Drioli, Italy
José Iván Escalante-García, México
Peter S. Hooda, United Kingdom
Mo Kim Hung, Malaysia
Kei-ichi Imamoto, Japan
Harald Justnes, Norway
Jacek Katzer, Poland
Satinder Brar Kaur, Canada
Said Kenai, Algeria
Rizwan Ahmad Khan, India
Jamal Khatib, United Kingdom
Olonade Kolawole, Nigeria
Ewa Korzeniewska, Poland
Vineet Kumar, India
Glenn M. Larkin, United States
Falk Liebner, Austria
Manuel Esteban Lucas Borja, Spain
Lucian Lucia, United States
Isabel Martins, Portugal
Ricardo Mateus, Portugal
Paula Milheiro-Oliveira, Portugal

Daniel V. Oliveira, Portugal
Solmoi Park, South Korea
Giovanni A. Plizzari, Italy
Alaa M. Rashad, Egypt
A.B. Danie Roy, India
Łukasz Sadowski, Poland
Nassim Sebaibi, France
Payam Shafigh, Malaysia
Ayyoob Sharifi, Japan
Raju Sharma, South Korea
Samendra P. Sherchan, United States
Caijun Shi, China
Jongsung Sim, South Korea
S.P. Singh, India

Shamsher Bahadur Singh, India
Kulvinder Singh, India
Shaleen Singhal, India
Zhong Tao, Australia
Bouziani Tayeb, Algeria
F. Pacheco Torgal, Portugal
Dai-Viet N. Vo, Vietnam
Ignacio Lombillo Vozmediano, Spain
Jan Vymazal, Czech Republic
Kejin Wang, United States
Frank Winnefeld, Switzerland
Xin Yang, China
Yifeng Zhang, Denmark
Wen Zhang, United States

Contents

Keynote Abstracts

xiv

Keynote Abstracts

F-EIR Conference 2021 on
Environment Concerns and its Remediation (ECR21)
Chandigarh, India, October 18–22, 2021

Microbial Electrochemistry for Environmental Engineering Applications

Yifeng Zhang[1]

ABSTRACT

Microbial electrochemical technologies and systems (METs), which employ microbes as catalysts to convert chemical energy stored in organic or inorganic matter into sustainable electricity or high-value products, is emerging as next generation technology of environmental electrochemistry. With extensive fundamental studies to understand the electron transfer mechanisms, microbiology, reactor architecture and materials involved, METs are becoming a versatile platform technology for various applications including wastewater treatment, environmental monitoring, chemical and biofuel production, resource recovery and waste remediation. This keynote aims to provide an updated overview of our key research activities in METs for environmental applications. Our research interests are microbial electrochemistry and biotechnology to support the 2nd, 6th, 7th, 12th, and 13th UN Sustainable Development Goals through sustainable water treatment, resource recovery, CO_2 capture and utilization, biosynthesis, and environmental bioremediation & monitoring. Our works were innovative and contributed considerably to improving the fundamental understanding of microbial electrochemistry and developing cutting-edge, efficient, and economic-affordable technical solutions in the Water-Food-Energy-Climate nexus for sustainable development and green transition. This way the wastewaters and greenhouse gases (e.g., CO_2), which otherwise are treated as pollutants undergoing energy and carbon-intensive treatment processes, are recovered and upcycled in a form of valuable bioproducts such as chemicals, biofuels, proteins, and energy.

Keywords: Microbial Electrochemistry, Water Treatment, Biosensor, Carbon Capture, Resource Recovery, Bioenergy.

---

[1] *Presenting & corresponding author.*
Associate Professor, Department of Environmental Engineering, Technical University of Denmark, DK-2800 Lyngby, Denmark. *E-mail:* yifz@env.dtu.dk

F-EIR Conference 2021 on
Environment Concerns and its Remediation (ECR21)
Chandigarh, India, October 18–22, 2021

Chemical Shrinkage of Pastes and Mortars Containing Limestone Fines

Jamal Khatib[1], Rawan Ramadan[2], Hassan Ghanem[2],
Adel ElKordi[2] and Oussama Baalbaki[2]

ABSTRACT

The purpose of this research is to investigate the effect of incorporating limestone fines (LF) on chemical shrinkage of cement pastes and mortars. Five paste mixes and 5 mortar mixes were employed. The cement was partially replaced with 0, 5, 10, 15 and 20% (by weight) LF. The water to binder ratio (w/b) was kept constant at 0.45 for all mixes. The sand to binder (s/b) ratio in the mortar mixes was 2. The tests conducted were; chemical shrinkage, compressive strength and density. Chemical shrinkage was tested each hour for the first 24 hrs, and thereafter each 2 days until a total period of 90 days. Furthermore, compressive strength and density were determined at 1 day, 7, 28 and 90 days of curing. The maximum chemical shrinkage of paste and mortar occurs in mixes containing between 10 and 15% LF. Beyond this level, the chemical shrinkage started to decrease. It was also noticed that compressive strength for pastes and mortars attained the highest value for mixes containing 10 and 15% LF. Density for pastes and mortars increased up to 15% LF followed by a decrease at 20% replacement level.

Keywords: *Limestone Fines, Chemical Shrinkage, Density, Compressive Strength, Pastes, Mortars.*

[1] *Presenting & corresponding author.*
Department of Civil Engineering, Beirut Arab University, P.O. Box 11 - 50 - 20 Riad El Solh 11072809, Beirut, Lebanon.
Faculty of Science and Engineering, University of Wolverhampton, Wolverhampton, WV1 1LY.
E-mail: j.khatib@bau.edu.lb
[2] Department of Civil Engineering, Beirut Arab University, P.O. Box 11 - 50 - 20 Riad El Solh 11072809, Beirut, Lebanon.

F-EIR Conference 2021 on
Environment Concerns and its Remediation (ECR21)
Chandigarh, India, October 18–22, 2021

Strategies for Reducing the Environmental Impact of Cementitious based Construction Materials

Luigi Coppola[1]

ABSTRACT

In recent years, the increase in the sustainability of building materials has become essential to preventing climate change caused by global warming. The use of construction materials defined as "sustainable" cannot be based exclusively on environmental parameters related to the production of mortars and concretes, such as the energy consumption, the natural raw materials consumption and the greenhouse gas emissions, but should also take into account the performance and the durability of the mixtures used during the construction of structural elements and secondary building components. The topic is particularly complex and requires a multidisciplinary approach that ranges from the development of new solutions for the production of binders and aggregates to the optimization of the production processes of portland cement clinker and to the improvement of waste management in order to transform slags and industrial by-products into a sustainable resource. Therefore, this keynote deals with the environmental impact of concrete industry and it highlights the best paths to improve the sustainability of cementitious materials, taking into account the reduction of environmental impact, the improvement of performances and the prolonging of service life of reinforced concrete structures. In particular, several strategies have been proposed: from the new technologies to improve the efficiency of cement plants (including the carbon capture, utilization and storage systems) to the partial replacement of Portland cement with supplementary cementitious materials, passing from the use of alternative "green" constituent (binders, aggregates and water) and the addition of admixtures for high durability/high performance concretes.

Keywords: Sustainability, Concrete, Environmental impact.

---

[1] *Presenting & corresponding author.*
Associate professor, Department of Engineering and Applied Sciences, University of Bergamo, Via Marconi 5, 24044 Dalmine (BG), Italy. *E-mail:* luigi.coppola@unibg.it

F-EIR Conference 2021 on
Environment Concerns and its Remediation (ECR21)
Chandigarh, India, October 18–22, 2021

Embedding Sustainability in Retrofit and Repair of Architectural Heritage

Marco Corradi[1] and Enea Mustafaraj[2]

ABSTRACT

The sustainability in remediation, retrofit and seismic upgrading of historic masonry structures is an important part of new sustainable construction methods. The largest part of existing building stock in many countries is made of stone and brick masonry. These structures have been often constructed many years ago using old technologies, methods, and often do not comply with the modern design standards. For many years, the solution adopted by governments and international bodies was to demolish these old masonry structures, suggesting the construction of new buildings, highly increasing water &energy consumption and waste, land take from agriculture, and put at risk the natural environment, landscapes and the fauna.

In the last decade, the increased awareness of preservation of historic structures because of the values they possess, has turned to development of sustainable strategies for retrofit and seismic upgrade of these structures. In this paper some rehabilitation techniques for old masonry and timber buildings are described and discussed. The selected methods aim at improving the structural performance, preserving the structures for future generations, having the least impact in altering the architectural and heritage values, as well as being sustainable.

Keywords: Historic Masonry Retrofit, Seismic Upgrade, Sustainable Solutions.

[1] *Presenting author.*

Department of Mechanical & Construction Engineering, Northumbria University, NE1 8ST Newcastle Upon Tyne, UK.

Department of Engineering, University of Perugia, 06125 Perugia, Italy.

E-mail: marco.corradi@northumbria.ac.uk

[2] Department of Civil Engineering, EPOKA University, 1039 Tirana, Albania.

E-mail: emustafaraj@epoka.edu.al

F-EIR Conference 2021 on
Environment Concerns and its Remediation (ECR21)
Chandigarh, India, October 18–22, 2021

Production of Green Artificial Aggregates Through Geopolymer Technology

Jian-Guo Dai[1]

ABSTRACT

Artificial aggregates can be produced from annually increasing industry by-products/ urban wastes through the sintering or cold bonding technology. Manufacturing artificial aggregates provides a one-stone two-birds solution to the above-mentioned problem. Production of green geopolymer lightweight aggregates (GLAs) is explored in this study through cold bonding and crushing techniques, which have a good prospect for massive industrial production. Geopolymer cubes were firstly produced using one-part geopolymer technology at ambient temperature and then crushed into both coarse and fine aggregates with angular shapes. Industry products like coal fly ash (FA) and ground granulated blast-furnace slag (GGBS) and anhydrous sodium metasilicate particles in industrial grades were utilized for producing the one-part geopolymer. Results showed that the produced GLAs with different mixes had a loose bulk density below 800 kg/m^3 and an apparent density of 1450–1750 kg/m^3 and could be used to produce high-performance geopolymer aggregate concrete (GAC) and fiber-reinforced cementitious composites due to the stronger aggregate-matrix interface and initial flaw effects. The findings are useful for future production and applications of GLAs in concrete industry.

Keywords: Geopolymer, Artificial Aggregates, Lightweight, Concrete, Composites.

[1] *Presenting & corresponding author.*
Ir. Professor, Department of Civil and Environmental Engineering, The Hong Kong Polytechnic University, Hong Kong, China. *E-mail:* cejgdai@polyu.edu.hk

Graphical Abstract

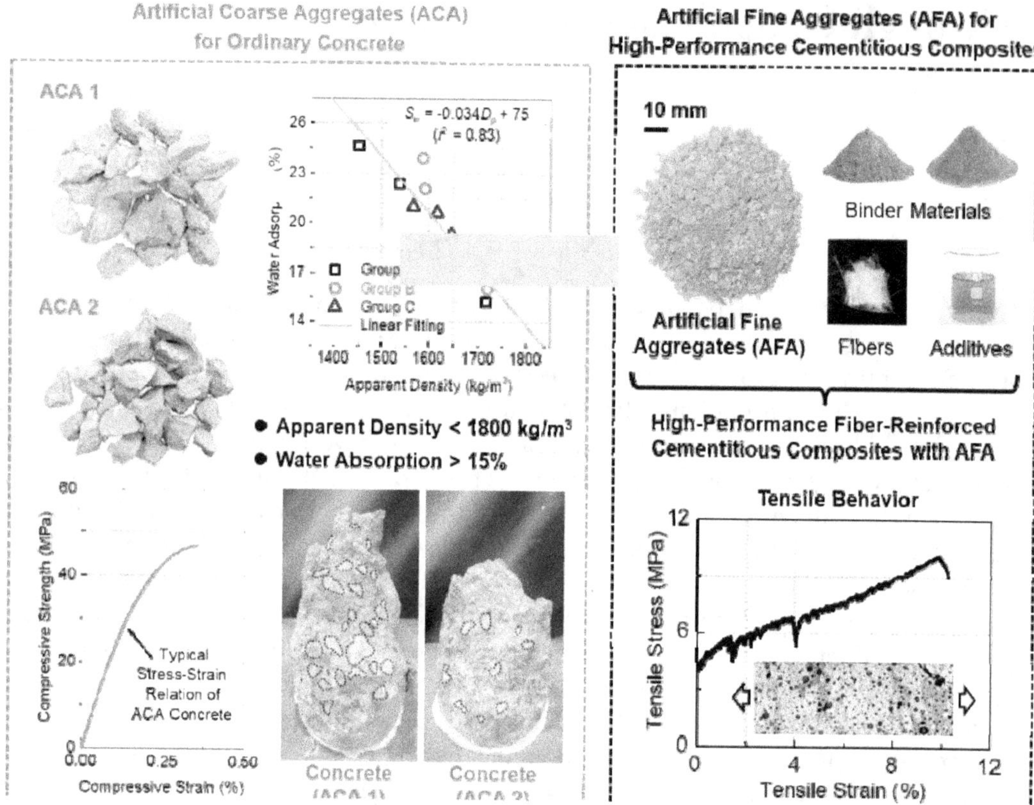

F-EIR Conference 2021 on
Environment Concerns and its Remediation (ECR21)
Chandigarh, India, October 18–22, 2021

MPCM-Based Porous Cementitious Composites for Enhanced Energy Efficiency of Smart Buildings

Mona Sam[1*], Liliya Dubyey[1#], Antonio Caggiano[2] and Eddie Koenders[1@]

ABSTRACT

This work reports a detailed experimental study that has the purpose of investigating the Thermal Energy Storage (TES) performance of cement pastes enhanced with a novel Microencapsulated Phase Change Materials (MPCMs). Three water-to-binder ratios and three MPCM volume fractions, leading to a total of nine different mixtures, were investigated at the *Institut für Werkstoffe im Bauwesen* of TU Darmstadt. Water-to-binder ratios of 0.33, 0.40 and 0.45 are considered, for the reference pastes, while a MPCM having a melting/solidification temperature range of 37°C and a latent storage capacity of 190 kJ/kg has been mixed in the MPCM-pastes. Thermal and mechanical tests, accompanied by SEM analyses, are performed to observe the effect MPCM have on the resulting TES and strength results. The experimental data have been analyzed to evaluate the corresponding temperature-based material parameters like specific heat, conductivity, or more in general, the energy storage capacity of these systems under transient heat transfer conditions.

Keywords: *Energy-Storage, Cement Pastes, Bio-based PCM, Thermal Energy Analysis, Heat Storage, Eco-friendly.*

[1] Institut für Werkstoffe im Bauwesen, TU Darmstadt, Germany.
E-mail: *sam@wib.tu-darmstadt.de; #dubyey@wib.tu-darmstadt.de; @koenders@wib.tu-darmstadt.de
[2] *Presenting & corresponding author.*
Antonio Caggiano, Adj. Prof. Dr. Dr.-Ing., Technische Universität Darmstadt, Institut für Werkstoffe im Bauwesen, Franziska-Braun-Straße 3, 64287 Darmstadt.
Universidad de Buenos Aires, FIUBA, LMNI, CONICET, Buenos Aires, Argentina.
E-mail: caggiano@wib.tu-darmstadt.de; caggiano@wib.tu-darmstadt.de; acaggiano@fi.uba.ar

Graphical Abstract

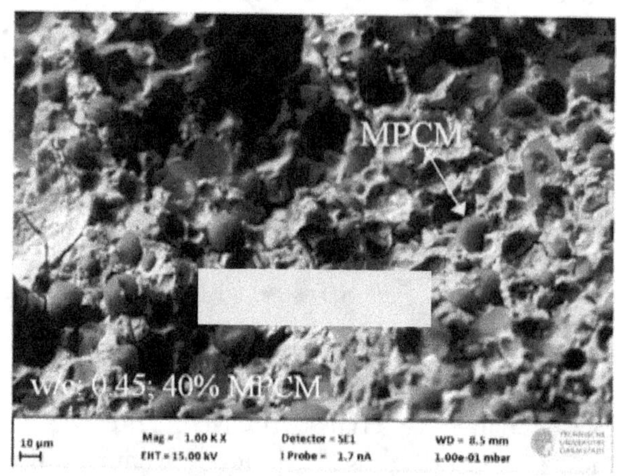

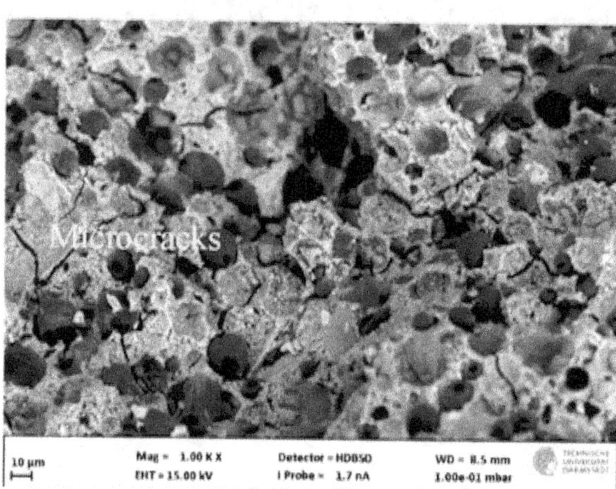

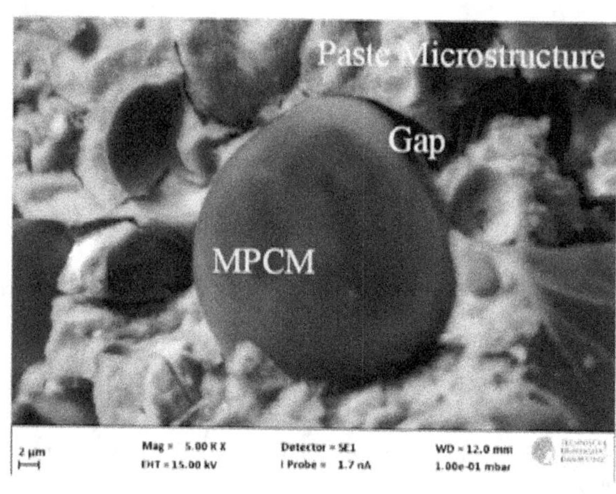

F-EIR Conference 2021 on
Environment Concerns and its Remediation (ECR21)
Chandigarh, India, October 18–22, 2021

Green Concrete with Clinker-Free Cement and Blast Furnace Slag Aggregate and Its Application to Concrete-Filled-Bamboo (CFB)

Kei-ichi Imamoto[1] and Chizuru Kiyohara[2]

ABSTRACT

This paper deals with the development of composite cement concrete-filled-bamboo (CFB) comprising pulverized waste plasterboard (P-WPB: G), fly ash (F), ground granulated blast-furnace slag (GGBFS: S) and bamboo. The fundamental properties of the F-S-G paste with regard to fluidity, drying shrinkage and compressive strength were tested, and the influence of the re-placement ratio of P-WPB was investigated. Using the optimum proportion of the F-S-G mixture, the properties of composite cement concrete (CCC), such as compressive strength, Young's modulus, shrinkage strain and air permeability were tested. Furthermore, a counter measurement with blast-furnace slag fine aggregate to decrease the autogenous shrinkage was conducted. Finally, author proposed concrete-filled-bamboo (CFB) and investigated the effect of reinforcement with bamboo throughout the bending strength of CFB.

- Composite cement concrete (CCC) exhibited sufficient compressive strength.
- Blast furnace slag aggregate (BFS) compensated shrinkage of CCC.
- CCC filled bamboo (CFB) exhibited sufficient bending strength capacity.

Keywords: Fly ash, GGBFS, Waste Plasterboard, Blast-Furnace Slag Fine Aggregate, Bamboo.

[1] *Presenting & corresponding author.*
Professor, Department of Architecture, Toyko University of Science, Japan.
E-mail: imamoto@rs.tus.ac.jp
[2] *Corresponding author.*
Research Assistant, Department of Architecture, Toyko University of Science, Japan.
E-mail: ckiyo@rs.tus.ac.jp

Graphical Abstract

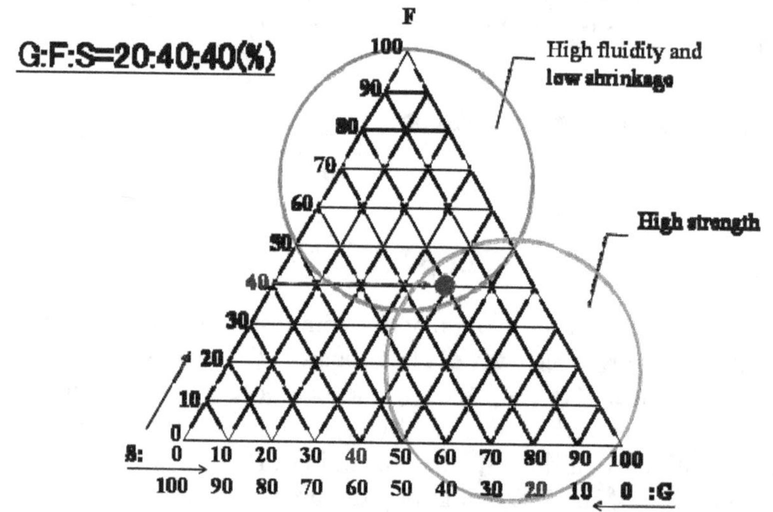

Bending Test of CFB

F-EIR Conference 2021 on
Environment Concerns and its Remediation (ECR21)
Chandigarh, India, October 18–22, 2021

Utilization of Fibres in Lightweight Cementitious Composite

Kim Hung Mo[1] and Geok Wen Leong[2]

ABSTRACT

With the increasing demand for sustainability in construction and materials, precast concrete component is one of the ways to reduce wastage and overall cost, as well as increasing construction speed, which can benefit from the application of lightweight concrete or cement-based materials. The incorporation of low-density hollow spheres such as perlite microsphere and fly ash cenosphere can produce sustainable lightweight cementitious composite (LCC). Good particle packing can be achieved by utilizing the fly ash cenosphere, which results in a LCC with high specific strength, exhibiting compressive strength of 40.8 MPa with an oven-dry density of 1406 kg/m^3. As lightweight cement-based materials typically are more brittle than normal concrete, fibre reinforcements are desired. The utilization of fibres, especially polyvinyl alcohol (PVA) fibre at 1.0% fibre volume was found to effectively improve the flexural strength and splitting tensile strength of the fly ash cenosphere LCC by about 100% and 65%, respectively. The introduction of alkali-resistant glass fibre textile mesh can also enhance the flexural performance of the LCC. As much as 4-folds in the flexural toughness was increased with the presence of 4 layers alkali-resistant glass fibre textile mesh in LCC containing 1.0% of PVA. Most importantly, the combination of the alkali-resistant glass fibre textile mesh and 1.0% PVA fibres demonstrated excellent synergistic effects, thus develops a significant enhancement in the ductility and tensile performance of thin plate LCC specimens. The experimental results show that the stiffness and tensile strain capacity of fiber reinforced LCC increased by about 85% and 35%, respectively, based on their tensile stress-strain response. This shows that the developed fibre reinforced LCC can potentially be incorporated and utilized in thin structure application.

Keywords: *Lightweight Cementitious Composite, Hollow Sphere Aggregates, Fibre, Fibre Textile Mesh.*

[1] *Presenting & corresponding author.*
Senior Lecturer, Department of Civil Engineering, Faculty of Engineering, Universiti Malaya, 50603 Kuala Lumpur, Malaysia. ***E-mail:*** khmo@um.edu.my
[2] PhD Candidate, Department of Civil Engineering, Faculty of Engineering, Universiti Malaya, 50603 Kuala Lumpur, Malaysia. ***E-mail:*** geokwen@um.edu.my

Multi-functional Behaviour of TiO_2 Cementitious Composites for Photocatalyst Air Cleaning and Energy Saving

Mona Sam[1*], Ronald Schiffer[1#], Maximilian Löher[1@], Rodrigo Gibilisco[2], Jörg Kleffmann[3], Eddie Koenders[1$] and Antonio Caggiano[4]

ABSTRACT

Air pollution is still among the biggest environmental health threats for humans in Europe. Traffic, industry and agriculture are the main responsible sources that are emitting air pollutants: i.e., nitrogen oxides (NO_x), sulfur dioxide (SO_2), carbon monoxide (CO) or particulate matter (PM). In this context, NOx are considered as one of the most harmful pollutants, which can have detrimental effects not only on humans and animals, but also on the environment. During the last two decades, novel multi-functional building materials have been designed by embedding photocatalysts employed to reduce (oxidize) NOx pollutants via photocatalytic reaction (PCR). Photocatalysts, like for example TiO_2, are using UV-light to convert adsorbed water and oxygen into highly reactive radicals (i.e., OH, O_2), which oxidize (clean) pollutants.

The present work reports recent developments and future trends on experimental and numerical research of employing a multi-functional (highly-porous) concrete foam, produced at the Institute of Construction and Building Materials of *Technische Univesität Darmstadt*, and enhanced with embedded TiO_2. A wide range of experimentally analyzed Thermal Energy Storage outputs, combined with the study of the photocatalytic activity, measured at the *Bergische Univesität Wuppertal*, are presented to demonstrate the promising multifunctionality of TiO_2 Foams.

Keywords: *Concrete Foam, NO× Cleaning, TiO₂ Powders, Multifunctional Composites, Photocatalysts.*

[1] Institut für Werkstoffe im Bauwesen, TU Darmstadt, Germany.
E-mail: *sam@wib.tu-darmstadt.de; #borges_schiffer@wib.tu-darmstadt.de; @loeher@wib.tu-darmstadt.de; $koenders@wib.tu-darmstadt.de
[2] INQUINOA (CONICET-UNT), Universidad Nacional de Tucumán, Argentina.
Universidad de Buenos Aires, FIUBA, LMNI, CONICET, Buenos Aires, Argentina.
E-mail: rgibilisco@fcq.unc.edu.ar
[3] Department of Chemistry and Biology, Bergische Universität Wuppertal, Germany.
E-mail: kleffman@uni-wuppertal.de
[4] *Presenting & corresponding author; Antonio Caggiano, Adj. Prof. Dr. Dr.-Ing., Technische Universität Darmstadt, Institut für Werkstoffe im Bauwesen, Franziska-Braun-Straße 3, 64287 Darmstadt.*
Institut für Werkstoffe im Bauwesen, TU Darmstadt, Germany.
INQUINOA (CONICET-UNT), Universidad Nacional de Tucumán, Argentina.
Universidad de Buenos Aires, FIUBA, LMNI, CONICET, Buenos Aires, Argentina.
E-mail: caggiano@wib.tu-darmstadt.de

Contributed Abstracts

F-EIR Conference 2021 on
Environment Concerns and its Remediation (ECR21)
Chandigarh, India, October 18–22, 2021

A Novel Plan to Mitigate the Effect of Flood, Drought and Salinity Using Interlinking of Canal and Rivers

Anant Patel[1], Neha Keriwala[2] and Upasana Panchal[3]

ABSTRACT

The rivers are not only life-line for humans, but also for wild-life. The rivers play a very important role within the lives of the people, because the Rivers are very much helpful in water supply, irrigation, water transportation, generation of electricity, tourism in addition as could be a source of livelihood for our increasing population. Most of the major cities are normally situated on the shores of holy rivers or near the coastal area. Proper management of river water is much needed. Agriculture activities are largely depends on the monsoon which is often undefined in nature specifically in India. Hence, there is a significant problem of lack of irrigation in an exceedingly region and water logging in others. To solve this problem it is essential to interlink all the rivers within the whole country to supply water to the deficit area. The approach is made store river water which discharge ample amount of water every year during monsoon season into the ocean. The main priority is given for interlinking of the Gujarat state major rivers to fulfill the needs of the areas which are facing the severe water scarcity such as North Gujarat, Saurashtra and Kutchh region of Gujarat State. This approach also helpful to those area where the effect of salinity is more. In recent years, there are lots of water crises and issues are reported across the world. So, it is right time to protect and conserve the surface water and also control the ground water being contaminated through the sea water intrusion for the future survival for the next generation. In this paper a sustainable resilient plan is suggested for river water interlinking, which is helpful in control of surface water storage and saline water intrusion into the ground water along the coastal area.

Keywords: *River Interlinking, Saline Water Intrusion, Canal Network, Drought & Flood, Water Resources Management.*

---

[1] *Presenting & corresponding author.*
Civil Engineering Department, Institute of Technology, Nirma University, Ahmedabad, India.
E-mail: anant.patel@nirmauni.ac.in
[2] Civil Engineering Department, Institute of Technology, Nirma University, Ahmedabad, India.
E-mail: keriwalaneha@gmail.com
[3] Town Planning & Valuation Department, UD&UHD, Government of Gujarat, Surat, India.
E-mail: upasanaoops5@gmail.com

F-EIR Conference 2021 on
Environment Concerns and its Remediation (ECR21)
Chandigarh, India, October 18–22, 2021

Ground Water Potential Zone Mapping Using Remote Sensing and GIS in Saurashtra Region of Gujarat, India

Anant Patel[1], Kinjal Zala[2] and Krupali Solanki[3]

ABSTRACT

Ground water is the water contained in the soil pores under the earth's crust. Water is collected in the aquifer underneath the soil, and the aquifer's holding ability is defined as Ground Water Potential. The challenges of ground water shortages and contamination have become worse in recent decades as a result of population development. Its proper management is required to alleviate these issues. Traditional approaches for groundwater conservation, such as groundwater based surveys, exploratory exploration, and geophysical techniques, are not only inefficient yet often time consuming. As a result, an advanced Remote Sensing and Geographic Information System (GIS) approach is used. Ground water management may be divided into distinct Potential Zones based on the amount of ground water accessible in a given region. Using Remote Sensing and GIS, any area's ground water potential zone can be calculated by combining tiff maps and assigning ranks and weightages to each parameter. Slope, drainage density, geomorphology, Lineament Density, Soil, Lithology, and LULC are the parameters used to determine the possible zone of ground water in any given region. This paper describes the work that went into determining the GW potential zone in Gujarat's Saurashtra district. Various interconnected maps of the Saurashtra area are created and combined in a GIS platform for this purpose. The area from which ground water can be readily accessed can be determined using the Ground Water Potential Zone chart, and this information can be used to dig wells and create hydrological structures.

Keywords: Ground Water, Potential Zone, RS& GIS, Water Level, LULC.

[1] *Presenting & corresponding author.*
Civil Engineering Department, Institute of Technology, Nirma University, Ahmedabad, India.
E-mail: anant.patel@nirmauni.ac.in
[2] Civil Engineering Department, Institute of Technology, Nirma University, Ahmedabad, India.
E-mail: 17bcl128@ nirmauni.ac.in
[3] Civil Engineering Department, Institute of Technology, Nirma University, Ahmedabad, India.
E-mail: 17bcl116@ nirmauni.ac.in

Environmental Remediation for Cementitious Systems Using Titania Nano-Composites

Mainak Ghosal[1] and Arun Kumar Chakraborty[2]

ABSTRACT

Growing environmental concerns and climate change effects have highlighted another major problem with concrete. Concrete's basic ingredients like cement, sand, rocks & water are all sourced from mother earth. With sand & stone mining ban and with water becoming progressively costly, what should be the new alternatives? For decades, major developments in cement-concrete performance were attained with application of micro-fine waste particles viz. fly ash, slag and silica fumes. But recently discovered exotic nano-scale materials, claims of alleviating the problem of natural materials scarcity and reducing all types of wastes including green-house gas (GHG) emissions. Nano titanium made from waste materials has excellent self-cleaning properties and is the most widely used nanomaterial in commercial-environmental applications.

Our paper reports the effect of adding hydrophobic 40 nm dia. nano-sized Titania (NT) particles by functionalized them to hydrophilic via. surface modification or wrapping by new generation pH neutral polymeric admixtures & added them to cement: sand ratio of 1:3 with a water-cement ratio of 0.4. Mechanical strength results of the cementitious composites were taken at all terms up to 365 Days (D). The optimized quantity of NT as found in cement composites are then added to a M-40 Grade concrete composites for compressive strength testing at both shorter & longer terms following Indian standard protocols. Simulating sea water behaviours in laboratory conditions – results of sulphate and chloride attacks on standard concrete at medium and longer terms is discussed. Results reveals that the application of NT in cementitious systems has an improved effect when compared with non-NT cementitious systems and also the former is more durable. Photocatalytic NT is also time durable and the micro-structural investigations disclose that crystallographic nature of NT, which is the main reason for their favourable mechanical impacts.

Keywords: *Cement; Composites, Concrete, Environment, Strength, Titania.*

[1] *Presenting & corresponding author.*
  Jt. Secretary, Coal Ash Institute of India, Kolkata, India.
  Ph.D. Scholar, Indian Institute of Engineering Science & Technology, India.
  E-mail: mainakghosal2010@gmail.com
[2] *Corresponding author.*
  Associate Professor, Civil Engineering Department, IIEST, India. *E-mail:* arun_besu_civil@yahoo.com

Fabrication of Chitosan-Hydroxyapatite Nano-adsorbent for Removal of Norfloxacin from Water

Arunima Nayak[1], Brij Bhushan[2] and Shreya Kotnala[3]

ABSTRACT

Norfloxacin is a widely used antibiotic but because of its poor absorption, is mainly excreted out. This accounts for its presence in municipal and domestic wastewater in significant concentrations. It is classified as an emerging pollutant whose presence in water bodies can lead to serious environmental disorders. Adsorption is widely recognized as a cleaner, environmentally friendly and cost-effective remediation option. Choice of the adsorbent dictates the efficiency of the treatment process. With an aim to ensure the sustainability of the treatment process and to enhance the physico-chemical features of the adsorbent, a biocompatible nanocomposite (HAP@CT) was fabricated from egg-shell waste derived hydroxyapatite and a natural polymer (chitosan). Characterization studies were conducted to establish the surface morphology, surface area, particle dimension, surface chemistry and crystallinity of HAP@CT. The removal performance of norfloxacin from simulated water was studied via batch adsorption experiments onto the fabricated adsorbent. Optimum parameters with respect to pH, contact time, temperature, adsorbate concentration and adsorbent dosage were determined for determining the conditions for achieving higher norfloxacin removal rates. Isotherm data was modelled using Langmuir and Freundlich while kinetic data was applied to pseudo-first order and second order models. Results revealed that norfloxacin showed high binding affinity to HAP@CT; with adsorption capacity increasing with increased adsorbent dosage. Maximum adsorption (92%) occurred at a dosage of 10 g/L and at a pH of 6.4. Kinetic studies revealed an equilibrium time of 20 mins. Adsorption data fitted well to Langmuir model (Langmuir adsorption capacity determined was 119 mg/g) while pseudo-second order kinetics best explained the kinetic data. Mechanism of binding of the adsorbate was determined to be physisorption as evident from FTIR studies, pH studies and isotherm modeling studies. The study established the successful fabrication of a biocompatible nanocomposite and its usefulness as an advanced adsorbent for fast and high removal of toxic contaminants like norfloxacin from wastewater.

Keywords: Norfloxacin, Hydroxyapatite, Nanocomposite, Adsorption, Isotherm, Kinetics.

[1,2,3]Department of Chemistry, Graphic Era University, Dehradun, India.
E-mail: [1]arunima_nayak@yahoo.com; [2]brijbhushaniitr@gmail.com; [3]shreya.kotnala001@gmail.com

F-EIR Conference 2021 on
Environment Concerns and its Remediation (ECR21)
Chandigarh, India, October 18–22, 2021

Estimating NDVI and LAI as a Precursor for Monitoring Air Pollution along the BBN Industrial Corridor of Himachal Pradesh, India

C. Prakasam[1], R. Aravinth[2] and B. Nagarajan[3]

ABSTRACT

The research conducted estimates the Leaf Area Index (LAI) through resolution multispectral images for Baddi Barotiwala Nalagarh. The LAI was estimated through high-resolution Multi-spectral imageries for the years (2001 to 2021). LAI analysis reveals that the majority of the study is covered under moderate to very high plant canopy density region. Analysis of the 2016 image revealed vegetation indices ranging between –0.9 to 2. Decrease of the forest vegetational quality was observed between 2001 to 2021 also causes the increase in the absorption of air pollutants while at the same time. In the year 2009 the RSPM of the areas has been registered at an average of 76 ug/m^3. Between 2010 to 2011 the average air quality registered at 106.67, 105.50, 83.33. The results indicate the air quality has been worsening through the years (2009 and 2020) RSPM of these regions has doubled increasing the toxicity of the pollutants. Analysis of the decadal change in NDVI and LAI shows the reduction of the annual forest coverage of the BBN industrial corridor coupled with the decrease in the air quality of the region. The study shows a clear relation between decreasing vegetational cover and the air quality of the region.

Keywords: *LAI, Regression Model, Remote Sensing, Air Pollution, BBN Industrial Corridor.*

[1] *Corresponding author.*
Department of Geography, School of Earth Science, Assam University (A Central University), India.
Department of Geography, School of Earth and Atmospheric Sciences, University of Madras, Chennai, India.
E-mail: cprakasam@gmail.com; c.prakasam@aus.ac.in

[2] *Presenting author.*
Department of Geography, School of Earth Science, Assam University (A Central University), India.
Department of Geography, School of Earth and Atmospheric Sciences, University of Madras, Chennai, India.

[3] National Centre of Geodesy, Indian Institute of Technology, Kanpur 208016, India.

F-EIR Conference 2021 on
Environment Concerns and its Remediation (ECR21)
Chandigarh, India, October 18–22, 2021

Effect of Rice Ash Husk on Permeation Characteristic of Cement Mortar

Arvind Vishavkarma[1] and Kizhakkumodom Venkatanarayanan Harish[2]

ABSTRACT

In this paper, the effect of rice husk ash (RHA) on chloride permeation, water absorption, and permeable pore space of cement mortar was investigated. It was observed that the permeability, water absorption, and permeable pore spaces of mortar mixtures having rice husk ash as partial replacement of cement decreased as the amount of rice husk ash increased while the compressive strength increased. The reduction in chloride permeation, water absorption, and permeable pore space was a range between ~3 to ~52, ~2 to 10%, and ~2% to 18% with 4% to 12% RHA dosages compared to without RHA mixture. In addition, linear co-relation exists between water absorption and permeable pore space with a coefficient of correlation of 0.93; quadratic polynomial co-relation exists between compressive strength and different permeation properties having a coefficient of correlation between ~0.60 to ~0.98.

Keywords: Chloride Permeation, Water Absorption, Permeable Pore Space, Linear and Quadratic Polynomial Co-Relation.

[1] *Presenting & corresponding author.*
Graduate Student, Department of Civil Engineering, Indian Institute of Technology, Kanpur 208016, India.
E-mail: arvind120491@gmail.com
[2] *Corresponding author.*
Assistant Professor, Department of Civil Engineering, Indian Institute of Technology, Kanpur 208016, India.
E-mail: kvharish@iitk.ac.in

F-EIR Conference 2021 on
Environment Concerns and its Remediation (ECR21)
Chandigarh, India, October 18–22, 2021

Landslide Susceptibility Analysis of Mandi District, Himachal Pradesh Using Hybrid Machine Learning Methods

Amol Sharma[1] and Chander Prakash[2]

ABSTRACT

Landslide susceptibility mapping is crucial step in comprehensive landslide risk management purpose of the present study is to analyse landslide susceptibility of Mandi district, Himachal Pradesh, India, based on optimum feature selection and hybrid integration of Shannon entropy (SE) model with random forest (RF) and support vector machine (SVM) models. An inventory of 1723 rainfall-induced landslides was generated and randomly selected for training (1199, 70%) and validation (524, 30%) purposes. A set of 14 relevant factors were selected and checked for multicollinearity. These factors were first ranked using Information Gain and Chi-square feature ranking algorithms. Then Wilcoxon Signed Rank Test and One-Sample T-Test were applied to check their statistical significance. An optimum subset of 11 landslide causative factors was then used for generating landslide susceptibility maps (LSM) using hybrid SE-RF and SE-SVM models. These LSM's were validated and compared using receiver operating characteristics (ROC) curves and performance matrices. The SE-RF performed better with training and validation accuracies of 96.93% and 88.94%, respectively, compared to the SE-SVM model with training and validation accuracies of 94.05% and 82.4%, respectively. The prediction matrices also confirmed that the SE-RF model was better and is recommended for landslide susceptibility analysis of similar mountainous regions worldwide.

Keywords: Landslide Susceptibility, Shannon Entropy, Random Forest, Support Vector Machine, Feature Selection, Performance Matrices.

[1] *Presenting & corresponding author.*
Department of Civil Engineering, NIT, Hamirpur, India.
E-mail: amol@nith.ac.in
[2] *Corresponding author.*
Department of Civil Engineering, NIT, Hamirpur, India.
E-mail: chandermanali@nth.ac.in

Study on Influence of Natural Aggregate Maximum Size on Compressive Strength of Polystyrene Aggregate Concrete of Structural Grade

Bogdan Rosca[1] and Adrian Alexandru Serbanoiu[2]

ABSTRACT

In the last decades, for concrete, expanded polystyrene (EPS) has been regarded as a new lightweight aggregate and the partial replacement of natural aggregates with EPS beads is a common technique to produce lightweight concretes. In the mainstream literature expanded polystyrene is considered an eco-friendly building material because it can contribute to the environment protection and conservation of the natural resources of aggregate. This paper comprises an experimental investigation on polystyrene aggregate concrete (PAC) having structural grade containing polystyrene beads in different percentages of replacement of the natural aggregate for which the maximum aggregate size (MAS) is varying. It is known in what extent the MAS influences the compressive strength of the ordinary concrete but is less known for concretes containing two types of aggregate with very different strength properties like PAC.

In this study, for coarse natural aggregate, the considered maximum sizes were 22, 16 and 8 mm, respectively. The MAS of natural aggregate was limited to 22 mm because the PAC is aimed for reinforced concrete members. The replacement volumes with EPS beads of the natural aggregate were in percentage of 25, 35 and 45%. For all mixes, the same particles distribution of aggregate was considered. Test results obtained on the influence of maximum size of natural aggregate on the properties like density and compressive strength for PAC containing EPS beads in different percentages are presented. The influence of MAS on the EPS beads distribution in PAC is presented, too. The results of the study indicate that the maximum aggregate size influences to some extent the strength of PAC when the natural aggregate particles are partly replaced with polystyrene EPS beads. For PAC containing beads in different replacement percentages of aggregate, the results indicate that increasing the maximum aggregate size leads to a decrease of the compressive strength.

Keywords: *Polystyrene Aggregate Concrete, Polystyrene Beads, Structural Strength, Maximum Aggregate Size.*

[1] *Presenting & corresponding author.*
 Assist Prof, Civil Engineering Faculty, Gh. Asachi Technical University, Iasi, Romania.
 E-mail: rosca.bogdan@ce.tuiasi.ro
[2] Civil Engineering Faculty, Gh. Asachi Technical University, Iasi, Romania. ***E-mail:*** serbanoiu.adrian@tuiasi.ro

F-EIR Conference 2021 on
Environment Concerns and its Remediation (ECR21)
Chandigarh, India, October 18–22, 2021

Land Use Land Cover Change Detection in Ladakh Using GIS and Remote Sensing Techniques

Mukesh Jamwal[1], Chander Prakash[2] and Amol Sharma[3]

ABSTRACT

Land use land cover (LULC) mapping and monitoring play a significant role in natural resource management and sustainable development. Remote sensing, coupled with geospatial tools, have excellent capabilities to study and assess the changes in land use and land cover on the regional scale. In the present study, land use and land cover changes of Ladakh region for the past two decades i.e. from 2000–2020, will be mapped and analysed using remote sensing satellite imagery and geospatial tools. The rapid growth of anthropogenic and tourism-related activities in the region and ongoing climate change has led to unplanned urbanisation and excess utilisation of natural resources which has put massive pressure on the already existing natural resources of the region. For sustainable development of the region, it is imperative to assess the LULC changes in recent years mainly focussing on built-up area, snow cover area, vegetation cover, water bodies and barren land. Landsat 7/8 images will be used for the study after correcting the same for radiometric and geometric disturbances. The image processing techniques of image enhancement, image classification and principle component analysis will be used for mapping and analysing the LULC changes in the past two decades in the region. High-resolution satellite data from google earth and field data from the area will be used for post-classification analysis and accuracy assessment. The study will be helpful in planning and decision making for sustainable development in Ladakh region.

Keywords: Land Use and Land Cover, Sustainable Development, Change Detection, Remote Sensing.

[1] *Presenting & corresponding author.*
Department of Civil Engineering, NIT, Hamirpur, India. *E-mail:* mukeshjamwal1377@gmail.com
[2] *Corresponding author.*
Department of Civil Engineering, NIT, Hamirpur, India. *E-mail:* chandermanali@nith.ac.in
[3] *Corresponding author.*
Department of Civil Engineering, NIT, Hamirpur, India. *E-mail:* amol@nith.ac.in

Experimental Study on Fly Ash and GGBS based Geopolymer Corbels

Sumanth Kumar Bandaru[1], K. Venkata Ramana[2] and D. Rama Seshu[3]

ABSTRACT

Cement is second most consumed material after water, due to phenomenal growth in infrastructure etc. leads to huge consumption of cement as building material. Geopolymer Concrete in this concern may prove to be a game-changing solution for the concrete industry as it has one of the potential to replace conventional cement in concrete. It also stresses on the utilization use of industrial wastes like fly ash, Ground granulated blast slag (GGBS) to be used as binder due to its chemical action with alkaline solutions to produce in-organic molecule. Fifteen reinforced geopolymer concrete corbels with M20 grade of concrete and different percentage of secondary reinforcement were cast and tested. The experimental shear strength results are related with few analytical and design codes. The test results showed that the shear capacity of geopolymer concrete underrates based on OPC concrete models of Corbels. The most analytical models are conservative in predicting shear capacity of GPC Corbels. However, Hagberg (1983) and Euro Code 2 (2004) predicts better shear capacity of GPC Corbels.

Keywords: *Geopolymer Concrete Corbels, Shear Strength, Empirical Approach, Shear Friction, STM.*

[1] *Presenting & corresponding author.*
Assistant. Professor, Civil Engineering Department, Kakatiya Institute of Technology and Science, Warangal, India. *E-mail:* bsk109@rediffmail.com

[2] Associate Professor, Department of Civil Engineering, Lakireddy Balireddy College of Engineering, Mylavaram, India.

[3] Professor, Civil Engineering Department, National Institute of Technology, Warangal, India.
E-mail: drseshu@nitw.ac.in

F-EIR Conference 2021 on
Environment Concerns and its Remediation (ECR21)
Chandigarh, India, October 18–22, 2021

Use of Particle Packing Methods for Development of Lime Fly Ash based Mortars for Repair of Heritage Structures

Degloorkar Nikhil Kumar[1] and Rathish Kumar Pancharathi[2]

ABSTRACT

Lime is a sustainable alternative to cement as it requires less energy for production and it absorbs CO_2 in the atmosphere for strength gain through carbonation. Non-hydraulic lime-based mortars have low strength and this can be enhanced using pozzolanic additions. However, densifying sand using particle packing theories in mortars for strength enhancement is one of the sustainable methods which is the main aspect of present study. In the current study, physico-chemical characteristics of two non-hydraulic lime binders Type-I and Type-II lime fly ash binders were prepared by replacing lime with fly ash from 0 to 30%. In addition, 5% gypsum was added to the binder mix for obtaining quicker setting. Two particle packing theories Modified Toufar Model (MTM) & J.D. Dewar Model (JDD) and Indian Standard Codes (IS: 650-2000 and IS: 383-2016) were considered for attaining different gradations of sand. Mix proportion 1:3 (Binder: Sand) lime-fly ash-based mortars were investigated for the flowability and compressive strength properties. The lowest minimum void ratio of 0.574 was attained for MTM proportioned sand. Compressive strengths of Type-I and Type-II lime fly ash-based mortars were highest for MTM particle packing sand with strengths 2.320 MPa and 0.738 MPa respectively after 28 days of curing. The compressive strengths were attained at 25% and 20% replacement levels of Type-I and Type-II limes with fly ash. The average flow values for Type-I and Type-II Lime fly ash-based mortars were obtained as 7.1 cm and 10.4 cm for MTM particle packing sand. From the study, it can be concluded that using Type-I lime fly ash-based mortar and MTM particle packing theory, repair mortars with reasonably good strength can be achieved.

Keywords: Lime, Sand, Fly ash, Particle Packing, Modified Toufar Method, JD Dewar Method, Compressive Strength.

---

[1] *Presenting & corresponding author.*
Research Scholar, Civil Engineering Department, National Institute of Technology Warangal, Warangal 506004, India. *E-mail:* nikhildegloorkar@gmail.com
[2] Professor, Civil Engineering Department, National Institute of Technology Warangal, Warangal 506004, India. *E-mail:* rateeshp@nitw.ac.in

Effect of Waste Sandstone Microfines on Mechanical Strength, Abrasion Resistance, and Permeability Properties of Concrete

Kusum Rathore[1], Vinay Agrawal[2] and Ravindra Nagar[3]

ABSTRACT

River sand is a non-renewable resource. The erosion of rocks obtains it, and this process takes thousands of years. The mining of river sand is increasing exponentially due to rapid economic growth and urbanization. Due to a shortage of river sand, low-quality river sand is being used in construction. Concrete can be made more resource-efficient and sustainable by using waste and recyclable materials for sand. Nowadays, manufactured sand (M-sand) is being utilized in construction to lessen the natural sand demand. Higher fines content, angular shape of particles, increased surface area, and rough texture are the key characteristics of M-Sand. The prime focus of this research article is to evaluate the potential of sand made from quarry waste of sandstone to be used as a 100 per cent replacement of river sand in concrete. Waste sandstone microfines were also used and replaced with fine aggregate from 5–25%, by weight with an increment of 5%. The generation of microfines is an unavoidable part of the sand manufacturing process. Microfines are particles of size less than 75 µm. These microfines are discarded either by sieving or washing, a waste of our valuable and non-renewable natural resources. This paper demonstrates the effect of the w/c ratio and sandstone microfine content on workability, mechanical strength, abrasion resistance, permeability, and ultrasonic pulse velocity of concrete. Test results showed that M-Sand and microfine content have a significant influence on mechanical strength, surface abrasion, and permeability of concrete. The strong interlocking of M-sand particles and the filler effect caused by microfines improved the mechanical strength and durability parameters of concrete. However, this positive impact of microfines in concrete is lowered down by the higher entrapped air at a low w/c ratio. This study establishes that using moderate amounts of sandstone microfines in M-sand can be a sustainable approach to solve the problem of natural sand shortage and reduce the loads from landfills.

Keywords: Sandstone, Microfines, M-Sand, Fine Aggregate, Abrasion Resistance.

---

[1] *Presenting & corresponding author.*
Research Scholar, Department of Civil Engineering, Malviya National Institute of Technology, Jaipur, India.
E-mail: 2017rce9049@mnit.ac.in
[2] Associate Professor, Department of Civil Engineering, Malviya National Institute of Technology, Jaipur, India.
E-mail: vagarwal.ce@mnit.ac.in
[3] Professor-HAG, Department of Civil Engineering, Malviya National Institute of Technology, Jaipur, India.

F-EIR Conference 2021 on
Environment Concerns and its Remediation (ECR21)
Chandigarh, India, October 18–22, 2021

Co-Utilization of Slag by Products from Steel Industries in Sustainable Concrete

Mohammed K.H. Radwan[2], Yi Zhi Hoo[2], Jerome Song Yeo[2], Chiu Chuen Onn[2] and Kim Hung Mo[1]

ABSTRACT

The global consumption of natural aggregates and cement is rising at an alarming rate due to the vast growing in construction industry. At the same time, the steel industries are generating large amounts of by-products in the form of electric arc furnace (EAF) slag, ladle surface (LF) slag, and ground granulated blast furnace slag (GGBS). These slag by-products can potentially be re-used and utilized as partial substitute of the common constituent materials in concrete such as the aggregates and cement. This paper thus investigates the influence of using the local EAF and LF slags as 40% aggregate replacement along with the GGBS at 25% cement replacement in concrete. The results demonstrated that improvement in the properties of concrete (such as compressive strength, water absorption, surface resistivity, and mass changes) can be achieved with EAF slag as partial coarse aggregate replacement, and these properties can be further enhanced with the use of GGBS. This points towards potential of the co-utilization of the EAF slag and GGBS in a sustainable concrete mixture to maximize the use of these industrial by-products.

Keywords: Electric Arc Furnace Slag, Ladle Furnace Slag, Ground Granulated Blast Furnace Slag, Waste Recycling, Sustainable Concrete.

[1] *Presenting & corresponding author.*
Department of Civil Engineering, Faculty of Engineering, Universiti Malaya, 50630 Kuala Lumpur, Malaysia.
E-mail: khmo@um.edu.my
[2] Department of Civil Engineering, Faculty of Engineering, Universiti Malaya, 50630 Kuala Lumpur, Malaysia.

Assessment of Environmental Flow Requirements through Rainfall Runoff Modelling for Hydropower Projects

C. Prakasam[1], R. Saravanan[2], M.K. Sharma[3] and Varinder S. Kanwar[4]

ABSTRACT

Catchment modelling using computer-based models has been identified as an effective tool in water resources planning and management. However, physically-based distributed computer models require either high-resolution data for a shorter period or large amounts of past data which are not readily available for some basins. Environmental flow is termed as the minimum required flow that has to be released on the downstream side of the hydropower project to maintain a healthy ecosystem. The NGT order defines the environmental flow as 15 % of the average lean season flow. In the Himalayan belt, there are many major and minor hydropower projects. The assessment of environmental flow requirements for these hydropower projects has many hiccups. One major disadvantage is the absence of the flow data or reliability of the data. To overcome this, the research work attempts at evaluating the inflow data through rainfall-runoff modeling. This methodology will be useful for all the major and minor hydropower projects with the data discrepancy. River basin planning and management are primarily based on the accurate assessment and prediction of catchment runoff. The rainfall, Soil, Evaporation, Drainage, Geomorphology, Geology data sets will be used in designing the model. The conceptual hydrological MIKE 11 NAM approach has been used for developing a runoff simulation model. It will be evaluated to result in the inflow of any river basin and hydropower project. The inflow data of the river basin will be used to calculate the environmental flow. The flow duration curve method has been chosen to do the same. The resultant will be compared with the National Green Tribunal of India order on environmental flow requirements and suitable interpretation will be made.

Keywords: *Rainfall Runoff Modelling, Environmental Flow, Conceptual Model, NGT, NAM Approach.*

[1] Corresponding author.
Associate Professor, Department of Geography, School of Earth Science, Assam University, (A Central University), Diphu Campus, India. ***E-mail:*** cprakasam@gmail.com
[2] *Presenting author.*
Research Scholar, Department of Civil Engineering, Chitkara University, India.
[3] *Scientist-E,* Environmental Hydrology Division, National Institute of Hydrology, India.
[4] Professor, Department of Civil Engineering, Chitkara University, India.

Performance Studies of Durability Properties of Recycled Aggregate Concrete

B. Suguna Rao[1], Chandrakanth[2] and S.M. Naik[3]

ABSTRACT

Industrialization and population explosion has increased the requirement from construction Industry. Continuous extraction and quarrying of natural aggregate (NA) for constructions are causing depletion of natural resources, causing a huge demand for natural resources. Industrialization also leads to the accumulation of large amount of construction and demolition waste being a part of dumping sites. Hence to reduce the facilitate the need of natural resource and also to reduce the impact on environment an effort is envisaged to utilize recycled aggregates in normal concrete by partially replacing coarse aggregates by recycled aggregates. Aggregates contributing to the 70% volume of concrete provides strength and stability to the concrete. After understanding the strength property of RAC, moisture movement of RAC is studied that affects the durability property of RAC. Strength of M60 grade of RAC concrete is studied by partially replacing natural aggregates by recycled aggregates. Moisture movement being one of the most important problems of durability is studied by considering normal concrete and RAC. RAC is tested for permeability, sorptivity and RCPT for normal concrete and RAC to understand the durability property of RAC and check its suitability in adverse situations.

Keywords: Construction and Demolition Waste (C&D), Normal Concrete (NC), Recycled Aggregate Concrete (RAC), Compressive Strength, Sorptivity, Permeability and RCPT Tests.

---

[1] *Presenting & corresponding author.*
Assistant Professor, Department of Civil Engineering, M.S. Ramaiah Institute of Technology, Bengaluru, India.
E-mail: suguna_rao@yahoo.com
[2] PG Student, Department of Civil Engineering, M.S. Ramaiah Institute of Technology, Bengaluru, India.
Professor, Department of Civil Engineering, M.S. Ramaiah Institute of Technology, Bengaluru, India.
E-mail: chandrakanthshamanna06@gmail.com
[3] PG Student, Department of Civil Engineering, M.S. Ramaiah Institute of Technology, Bengaluru, India.
E-mail: srikanth_naik@yahoo.com

F-EIR Conference 2021 on
Environment Concerns and its Remediation (ECR21)
Chandigarh, India, October 18–22, 2021

Planning Strategies to Improve Deteriorating Living Environment of Hill Towns: A Case of Dharamshala

Adite Dhadwal[1] and Ashwani Kumar[2]

ABSTRACT

Hills are losing their true identity due to the overpowering of concrete world. The amalgamation of its local elements with nature are getting deteriorate. Hill stations are famous for its pattern of development and sustainable growth where every element responds at its best with the nature around and the geological conditions. Today due to the rapid urbanizations people's interest have shifted from basic vernacular practices to luxurious modern practices which are ultimately causing disturbance within its extremities in various aspects. Hill stations are facing the fade away of their vegetation cover, traditional architecture and also derailing of natural drains. The locally available materials like stone, bamboo, slate, mud and wood are no longer put in good quantity as before which are strong to face harsh weather, geological conditions and also economically less costly. Over excavations are penetrating deep in the stability of the geological conditions thereby increasing its tendency to be unstable and unfit to reside upon. Planners and architects need to understand the alarming situation of the deteriorating of living environment in hills and so must put forward the designs and policies that make the town resilient with efficiency to work sustainably. We all need to be efficient enough to plan and strategies the future development before our thirst for luxury declines the quality of hills and make it worthless for living. Similar is the condition we are noticing in Dharamshala, where the unplanned world of concrete with rapid urbanization is eating up the nature, its breathing organs and making it a congested, unhygienic and worthless of living in near future.

This paper will emphasize on various issues hills town are facing along with some strategies with which we planners can regulate the development to grow sustainably. Hills are the gift of Himalayas we have with us. Its preservation is as important as its use to relish it for years. The paper will be compiled on the basis of secondary data and primary reconnaissance survey of the case study area.

Keywords: Hill Stations, Sustainability, Geological Disaster, Vernacular Architecture, Resilient Town.

---

[1] *Presenting & corresponding author.*
Assistant Professor, University Institute of Architecture, Chandigarh University, Mohali 140413, India. *E-mail:* aditedhadwal04@gmail.com
[2] Associate Professor, Department of Architecture and Planning, Department of Architecture and Planning, Jaipur 302017, India. *E-mail:* akumar.arch@mnit.ac.in

Assessing Environmental Vulnerability to Increased Occurrences of Avalanche: A District Level Assessment in Indian Western Himalayan Region

Akshay Singhal[1], Rajnish Kumar[2] and Sanjeev Kumar Jha[3]

ABSTRACT

Indian Western Himalayan Region (IWHR) is highly sensitive to changing weather patterns. Snow scientists all over the world are worried of the accelerating impacts that increased warming and rise in untimely precipitation events may have over the region. One such consequence of climate change is the increased occurrences of avalanche witnessed in the region over the past few years. Snow avalanches are generally influenced by variability of snow and weather conditions of the region alongside their interaction with complicated mountainous topography. However, due to increased anthropogenic activities in the IWHR, triggering of avalanches are not limited to natural reasons alone. A significant rise in native population and tourists, rapid building of roads and tunnels are also among the few causes which are exerting pressure on land use. Moreover, occurrences of moderate to serious avalanches can cause huge loss to life and property. Being spontaneous in nature, the local environment of IWHR remains highly vulnerable to avalanches. There is lack of studies which evaluates the overall risk potential of avalanche prone regions, especially in the Indian context. Most studies focus on the changing weather patterns to evaluate the causes avalanche. In this study, we adopt the framework of Intergovernmental Panel on Climate Change (IPCC) to conduct a vulnerability assessment in the avalanche prone districts of the IWHR with the focus on a composite evaluation including indicators of weather, complex topography, vegetation and local socio-economic conditions aiming to formulating an Avalanche Vulnerability Index (AVI). Each indicator is allotted dimensions of exposure, sensitivity and adaptive capacity based on its nature of function. The indicators are then assigned appropriate weights and subsequently aggregated to form the AVI. Based on the severity of avalanche hazard, districts are classified as 'Low', 'Moderate', 'High' and 'Extreme'. Results are used to produce Avalanche Vulnerability Maps (AVMs) for each of the avalanche prone districts of the region. Further, suitable adaptation measures are suggested to mitigate future losses. Overall, this study can be utilized by government agencies as a guide to plan and prepare action plans among districts of IWHR identified as vulnerable to avalanches.

Keywords: *Vulnerability, Avalanche, Western Himalayas, Exposure, Sensitivity, Adaptive Capacity.*

[1] *Presenting & corresponding author.*
  Indian Institute of Science Education and Research Bhopal, India. *E-mail:* akshay19@iiserb.ac.in
[2] Indian Institute of Science Education and Research Bhopal, India. *E-mail:* rajnish16@iiserb.ac.in
[3] *Corresponding author.*
  Indian Institute of Science Education and Research Bhopal, India. *E-mail:* sanjeevj@iiserb.ac.in

Geotechnical Characterization of Expanded Polystyrene (EPS) Beads with Industrial Waste and Its Utilization in Flexible Pavement

Rohan Deshmukh[1], Saivignesh Iyer[2] and Prathamesh Bhangare[3]

ABSTRACT

Conventional materials are used extensively for construction of road pavements which can one day lead to its' depletion. Expanded polystyrene (EPS) beads which are lightweight cellular plastic polymeric material is an excellent alternative material along with industrial wastes like fly ash and bottom ash that can be for the construction of flexible road pavements. This study describes the engineering behavior of EPS beads as a geo-material of different combinations with fly ash (FA), bottom ash (BA) and expansive soils (ES). Laboratory tests including compaction test were performed on samples of various proportions of EPS beads of EPS beads + Fly ash (EFA), EPS beads + Bottom ash (EBA), and EPS beads + Expansive soil (EES) to calculate the optimum value of addition of EPS beads to EFA, EBA and EES samples that would provide the highest strength to it. California Bearing Ratio (CBR) value and other strength parameters were calculated for optimal mixes of EFA, EBA and EES samples. Optimum size of EPS beads size was considered as 1 mm. Results of laboratory study suggest that addition of 0.5% EPS beads by weight to fly ash, bottom ash and expansive soil enhances its shear and bearing characteristics. Numerical simulations were carried out on PLAXIS 2D for optimal mixes of EFA, EBA and EES pavement blocks to check its behavior and suitability as a replacement of granular sub-base (GSB) layer materials in flexible pavement. Pavement design charts were also prepared to present a comparison between conventional and EPS composite pavements. The results of numerical analysis suggest that EFA and EBA can replace maximum 21.33% and an average of 12.25% of total thickness of conventional GSB layer. Cost analysis of overall road pavement showed that EFA, EBA and EES can reduce maximum overall costs up to 12.87%, 12.09% and 10.56% respectively when replaced with conventional GSB layer materials of the road pavement.

Keywords: EPS Beads, Fly Ash, Bottom Ash, CBR Value, PLAXIS 2D.

[1] *Corresponding author.*
Assistant Professor, Department of Civil Engineering, Terna Engineering College, Navi Mumbai, India.
E-mail: rohandeshmukh520@gmail.com
[2] *Presenting author.* Student, Department of Civil Engineering, Terna Engineering College, Navi Mumbai, India.
E-mail: vigneshyr505@gmail.com
[3] Student, Department of Civil Engineering, Terna Engineering College, Navi Mumbai, India.
E-mail: prathambhangare1234@gmail.com

F-EIR Conference 2021 on
Environment Concerns and its Remediation (ECR21)
Chandigarh, India, October 18–22, 2021

Impacts of Polyvinyl Alcohol and Basalt Fibres on Green Fly Ash Cenosphere Lightweight Cementitious Composite

Geok Wen Leong[2], Emeer Haziq H. Pahdili[2], Kim Hung Mo[1] and Zainah Ibrahim[2]

ABSTRACT

In this research, a green lightweight cementitious composite (LCC) with high specific strength, incorporating fly ash cenosphere (FC) as silica sand replacement, as well as 50% GGBS as the binder material, is investigated. This study examines the influence of polyvinyl alcohol (PVA) and basalt fibres, along with various fibre contents from 0.1% to 0.75%, on the green FC-LCC. The consistency, density, compressive and flexural strengths of the fibre reinforced LCC were assessed. The compressive strength of the fibre reinforced LCC is in the range of 41.7–50.3 MPa at densities of about 1500–1580 kg/m^3. The LCC with inclusion of up to 0.75% PVA fibres generally performs satisfactorily in terms of the mechanical strengths, and higher PVA fibre content can be introduced. For the case of LCC reinforced with basalt fibres, however, the fibre content is recommended to limit to 0.5% as excessive fibre content can cause difficulty in compaction and reduce the mechanical strengths.

Keywords: *Fly Ash Cenosphere, Lightweight Cementitious Composite, PVA Fibre, Basalt Fibre.*

[1] *Presenting & corresponding author.*
Department of Civil Engineering, Faculty of Engineering, Universiti Malaya, 50603 Kuala Lumpur, Malaysia.
E-mail: khmo@um.edu.my
[2] Department of Civil Engineering, Faculty of Engineering, Universiti Malaya, 50603 Kuala Lumpur, Malaysia.

F-EIR Conference 2021 on
Environment Concerns and its Remediation (ECR21)
Chandigarh, India, October 18–22, 2021

Composite Asphalt Binder Modification with Waste Non-Tire Automotive Rubber and Pyrolytic Oil

Ankush Kumar[1], Rajan Choudhary[2] and Abhinay Kumar[3]

ABSTRACT

Ethylene-propylene-diene monomer (EPDM) rubber is a synthetic rubber which shares about one-half of the total consumption of non-tire automotive rubber parts such as weather strips, window seals, hoses, *etc.* Pyrolysis is gaining attention as a thermochemical route for the effective and sustainable treatment of end-of-life tires. Tire pyrolytic oil (TPO) is one of the principal products generated from the pyrolysis of scrap tires. This study evaluated the use of TPO and EPDM waste rubber in composite asphalt binder modification through rheological characterization. In addition, the study also compared two processes (with and without pre-treatment/premixing of TPO and EPDM) for the preparation of the composite modified asphalt binders. The high temperature rheological characterization of binders was executed by the Superpave rutting criterion, failure temperatures, and multiple stress creep recovery (MSCR) test. Frequency sweeps were conducted at multiple temperature to study the changes in complex modulus and phase angle master curves. Also, the frequency sweep data was employed for analyzing the viscoelastic behavior of binders through storage and loss moduli crossover response. Furthermore, rheological tests including Superpave fatigue parameter, linear amplitude sweep (LAS), and binder yield energy test (BYET) were conducted at intermediate temperatures on the long-term aged binders. The observed results suggested an improvement in the asphalt binder performance with the composite modification through waste EPDM rubber and TPO in comparison to the single modification with EPDM and TPO alone. Pre-mixing of TPO and EPDM before adding to asphalt binder proved to be a better approach for preparing composite modified binder with EPDM rubber and TPO and yielded better high- and intermediate-temperature performance.

Keywords: *Composite Asphalt Modification, Pyrolytic Oil, EPDM Rubber, Asphalt, Waste Tire, Pre-treatment.*

---

[1] *Presenting author.*
 PhD Research Scholar, Department of Civil Engineering, IIT Guwahati, India.
 E-mail: ankus174104035@iitg.ac.in
[2] *Corresponding author.*
 Professor, Department of Civil Engineering, IIT Guwahati, India. *E-mail:* rajandce@iitg.ac.in
[3] PhD Research Scholar, Department of Civil Engineering, IIT Guwahati, India.
 E-mail: abhinay.kumar@iitg.ac.in

F-EIR Conference 2021 on
Environment Concerns and its Remediation (ECR21)
Chandigarh, India, October 18–22, 2021

Use of Response Surface Methodology for Optimization of Fluoride Removal Efficiency by Adsorption on Black Mustard Husk Ash

Akash Sitaram Jadhav[1] and Madhukar Vinayak Jadhav[2]

ABSTRACT

Fluoride removal techniques have been studied for several years. In recent years various fluoride removal techniques have been developed in order to curb the health hazards from fluoride. Adsorption is the most widely used and researched method due to its advantages like economy, highly efficient, simplicity of application etc., over all other available defluoridation methods. This paper deals with the study of adsorption of fluoride using black mustard husk ash as an adsorbing material. The process parameters like pH, adsorbent dose, contact time and stirring rate were optimized using a three-level, four-factor, Box–Behnken design. The regression analysis of the experimental data acquired from 27 runs aided to develop a second order mathematical model in order to predict the fluoride removal efficiency. Using the response optimization, the optimum fluoride removal efficiency was found to be around 87%. The surface response graphs for the different process parameters were plotted and analyzed for studying their effect on adsorption phenomenon. The p-value obtained from the Analysis of Variance model suggested that the model was significant (p-value less than 0.005). The optimum values of pH, contact time, stirring rate and adsorbent dose were found to be 5, 75 min, 200 rpm and 2 g respectively. The predicted and adjusted R^2 values were found to be 0.9177 and 0.9264 respectively. Adsorption models like Freundlich and Langmuir were used to validate the results. Langmuir model was seen best fiiting with the results having a R^2 value equal to 0.9029. Pseudo first order kinetics and pseudo second order kinetic modelling was studied. The kinetic models showed the influence of physisorption as well as some amount of chemisorption in the adsorption process. From the results obtained it was concluded that the black mustard husk can be efficiently utilized for adsorption of fluoride from water.

Keywords: *Adsorption, Black Mustard, Box-Behnken Design, Fluoride, Response Surface Methodology.*

---

[1] *Presenting & corresponding author.*
Research Scholar, Department of Civil Engineering, SRES Sanjivani College of Engineering, Kopargaon, Savitribai Phule Pune University, Pune, India. *E-mail:* er.asjadhav@gmail.com
[2] Professor, Department of Civil Engineering, SRES Sanjivani College of Engineering, Kopargaon, Savitribai Phule Pune University, Pune, India. *E-mail:* jadhavmadhukarcivil@sanjivani.org.in

F-EIR Conference 2021 on
Environment Concerns and its Remediation (ECR21)
Chandigarh, India, October 18–22, 2021

New Age Ternary Blended Cement Free Polymer Concrete: Manufacturing, Characterization and Evaluation of Mechanical Properties

Sudheer Ponnada[1], V.R. Sankar Cheela[2] and S.S.S.V. Gopala Raju[3]

ABSTRACT

This investigation gives a view on new age cement free polymer concrete made with coarse aggregate fully replaced with contrived aggregates (i.e., made with fly ash and thermosetting polymers) and fine aggregate also fully replaced with quarry dust. The compressive strength of the above new age cement free polymer concrete specimens at an age of 3, 7 and 14 days as confirming to IS standards were determined to study the early age strength with non-conventional ingredients. This work mainly throws light on the compressive strength and applicability of the above-mentioned new age cement free polymer concrete in construction industry with non-conventional ingredients. In this present investigation, thermosetting polymer as a major substance is used as a binder for the casting of contrived aggregates and quarry dust as fine aggregate for creating cement free concrete with fly ash, ground granulated blast furnace slag, NaOH and Na_2SiO_3 as normal geo polymer. In the early study of corresponding author study the different percentages of thermo setting polymer by weight of fly ash is weighed and mixed with fly ash and as per the results an optimum mix was achieved to manufacture contrived aggregates with indigenous equipment and same were tested for all the properties of natural aggregates confirming to IS standards. In this current study, to investigate the suitability of contrived aggregate and quarry dust as non-conventional ingredients in cement free concrete with geo polymer. Specimens are laid by fully replacing the natural aggregates with non-conventional aggregates. Compression test was performed on the same new age cement free concrete samples at above mentioned days for their suitability in contruction industry. The microstructure morphology was conducted using scanning electron microscopic image analysis and energy dispersive X-ray spectroscopy. All results were found satisfactory to use this innovative new age cement free concrete in conventional concrete practice.

Keywords: *Quarry Dust, Contrived Aggregate, Fly Ash, Ground Granulated Blast Furnace Slag, Thermo Setting and Geo Polymers, Compressive Strength.*

[1] *Presenting & corresponding author.*
Department of Civil Engineering, MVGR College of Engineering (A), Vizianagaram 535005, India.
E-mail: sudheer.ponnada@gmail.com
[2] *Corresponding author.*
Department of Civil Engineering, Indian Institute of Technology Kharagpur 721302, India.
[3] *Corresponding author.*
Department of Civil Engineering, Rajiv Gandhi University of Knowledge Technologies, Nuzvid 521202, India.

F-EIR Conference 2021 on
Environment Concerns and its Remediation (ECR21)
Chandigarh, India, October 18–22, 2021

High Temperature Impact on Calcined Clay-Limestone Cement Concrete (LC^3)

S.M. Gunjal[1] and B. Kondraivendhan[2]

ABSTRACT

A lot of CO_2 is released into the atmosphere during the manufacturing process of cement, which causes a global warming problem. Calcined clay-limestone cement concrete (LC^3) can be used to partially solve this problem. Limestone cement concrete is produced from calcined clay-limestone cement which is an advanced ternary mixed cement formed by the combination of 30 percent low-grade calcined clay, 15 percent limestone powder, and 5 percent gypsum and 50 percent cement clinkers. That means 50 percent of cement clinkers are to be replaced by additional cementitious materials which are beneficial for reducing CO_2 emission at the time of production of cement. In the present study, calcined clay-limestone cement is used to prepared concrete and compare it with concrete using ordinary portland cement (OPC) of grade 53. The physical and chemical characteristics of prepared cement samples were carried out and compared with OPC. M30 grade of concrete was prepared by calcined clay-limestone cement; mechanical properties such as compressive strength (CS) test are carried out on prepared concrete. After the age of 28 days, M30 grade of concrete was exposed to 100°C, 200°C, 400°C, 600°C, and 800°C temperature, surface texture, weight loss, compressive strength test were carried on the cubes of size 150 × 150 × 150 mm^3. The SEM analyses were carried out over 100°C and 600°C temperature on OPC and calcined clay-limestone cement concrete. The result shows that the compressive strength reduces with an increase in exposed temperature. The compressive strength decreases after 200°C and significant losses occur after 400°C. When the temperature reached 400°C, noticeable minute cracks started to appear on the surface of the cube specimens.

Keywords: High Temperature, Calcined Clay, Limestone, Weight Loss, Compressive Strength.

---

[1] *Presenting & corresponding author.*
Department of Structural, Sanjivani College of Engineering, Savitribai Phule Pune University, Kopargaon 423603, India. *E-mail:* gunjalsachin20@gmail.com
[2] *Corresponding author.*
Civil Engineering Department, S.V. National Institute of Technology, Surat 395007, India.
E-mail: kondraivendhan78@gmail.com

F-EIR Conference 2021 on
Environment Concerns and its Remediation (ECR21)
Chandigarh, India, October 18–22, 2021

Thermosetting Polymer, Quarry Dust and Fly Ash based Mortar as a Construction Material for Coastal Structures: Assessment of Fresh and Hardened Concrete Properties

Sudheer Ponnada[1], V.R. Sankar Cheela[2], M.G. Muni Reddy[3] and S. Adiseshu[3]

ABSTRACT

Present investigation is to produce cementless concrete utilized in coastal zones. Same was prepared with fly ash, quarry dust, thermosetting polymer and with conventional coarse aggregate in the present study. Also examine its mechanical properties and compare with conventional concrete. The experimental specimens with fly ash, quarry dust, and thermosetting polymer have been prepared by means of an advanced process merger with conventional coarse aggregate. Thermosetting polymer-based hybrid mortar categorized by different contents of normalized fly ash and quarry dust by a homogeneous distribution of the polymer have been accomplished. Once hardened the above hybrid mixture with new composite materials shows improved compressive and flexural strength in respect to both the fly ash and quarry dust pastes since the polymer provides a more cohesive microstructure, with a reduced number of micro voids and cracks at an early age of 3 hrs. at 1000C and 24 hrs. at 500C. The microstructural characterization allows pointing out the presence of an ITZ like that detected in cement mortars. A correlation between microstructural features and mechanical properties of the concrete made with advanced mortar has also been studied.

Keywords: Fly ash, Quarry Dust, Thermosetting Polymer, Compressive Strength, Tensile and Flexural Strength.

[1] *Presenting & corresponding author.*
Department of Civil Engineering, MVGR College of Engineering (A), Vizianagaram 535005, India.
E-mail: sudheer.ponnada@gmail.com
[2] *Corresponding author.*
Department of Civil Engineering, Indian Institute of Technology Kharagpur 721302, India.
[3] *Corresponding author.*
Department of Civil Engineering, Andhra University College of Engineering (A), Visakhapatnam 530003, India.

F-EIR Conference 2021 on
Environment Concerns and its Remediation (ECR21)
Chandigarh, India, October 18–22, 2021

Torsion Behavior of Strengthened RC Beams by Ferrocement

Rajesh S. Rajguru[1] and Manish Patkar[2]

ABSTRACT

Now-a-days, construction industry is one of the most over loaded industry and is responsible for carbon emission the process of enhancing the life of existing structures reduces the impact on the environment. In this regards, external strengthening of concrete structures is necessary under repair and rehabilitation of structures. This will increase the load carrying capacity and life span of concrete structures. In past decade fiber reinforced polymer (FRP) has been widely used as externally strengthening material due to its high strength and light weight. However, ferrocement is one of the suitable alternatives over costly FRP strengthening technique. Ferrocement is thin composite material having better crack resistance ability, ductility, good impact resistance and can be easily wrapped over beam surfaces. Very less research work is available on torsional behavior of concrete beam strengthened by ferrocement due to its complex loading. As practically beam and slab are cast monolithically, strengthening of beam in U-shape is more practical than strengthening all sides. This paper represents an experimental study conducted on reinforced concrete beam strengthened externally by U-shaped ferrocement with different strengths of mortar. This paper also addresses the torsional capacity and angle of twist of reinforced concrete beams. The experimental study was conducted on total 10 number of reinforced concrete (RC) beams of 150 mm × 200 mm × 1500 mm in size. The main purpose of study was to check the effect of different strength of mortar on ferrocement strengthening system under torsion keeping layer of wire mesh and thickness constant. Different strength of mortar mix, fabricated by varying water/cement ratio of 0.400–0.475 at stepped increment of 0.025. Result shows very less improvement in ultimate torsional strength and average 10.33% increment in angle of twist over normal beams.

Keywords: *Ferrocement, Reinforced Concrete, Torsions, Strengthening, Angle of Twist.*

---

[1] *Presenting & corresponding author.*
Research Scholar, Civil Engineering Department, Oriental University, Indore, India.
E-mail: rajgururs25@gmail.com
[2] Professor, Civil Engineering Department, Oriental University, Indore, India.
E-mail: manish.patkar@gmail.com

Photocatalytic Degradation of a Macrolide Using Fe Doped TiO$_2$

Hema Setia[1], R.K. Wanchoo[2] and Amrit Pal Toor[3]

ABSTRACT

The occurrence of emerging contaminants in the aquatic environment has become a worldwide concern as they have the potential to induce harmful effects on human health and ecosystems even at trace concentrations. Nanotechnology offers ample opportunities to develop newer treatment process that can help mitigate the harmful effects of such contaminants. This study reports the photocatalytic degradation of azithromycin, a macrolide antibiotic using titanium dioxide (TiO$_2$) and Fe doped TiO$_2$. 2 wt% Fe doped TiO$_2$ with a bandgap energy of 2.69 eV allowed for 73.92% degradation of 25 mg/L azithromycin in 210 min under solar irradiation while TiO$_2$ allowed 63.96% degradation under ultra violet (UV) irradiation in 120 min. Disk diffusion essay was done to evaluate the toxicity of end products of degradation.

Keywords: Photocatalysis, Degradation, Macrolide, Azithromycin, TiO$_2$.

[1] *Presenting & corresponding author.*
Biotechnology, University Institute of Engineering and Technology, Panjab University, Chandigarh, India.
E-mail: hemasetia@pu.ac.in
[2] Dr SSB University Institute of Chemical Engineering and Technology, Panjab University, Chandigarh, India.
E-mail: wanchoo@pu.ac.in
[3] Dr SSB University Institute of Chemical Engineering and Technology, Panjab University, Chandigarh, India.
E-mail: aptoor@yahoo.com

F-EIR Conference 2021 on
Environment Concerns and its Remediation (ECR21)
Chandigarh, India, October 18–22, 2021

Study of Pozzolanic Activation of Various Industrial Wastes and its Suitability in the Development of Geopolymer Matrix

Santosh R. Nawale[1] and Subhash V. Patankar[2]

ABSTRACT

Geopolymer concrete is a new construction material in which cement is fully substituted by industrial waste containing high amounts of silica and alumina, which is then activated by alkaline solutions. In the present investigation, various industrial wastes such as fly ash, GGBS, Calcined clay, and micro-silica were used as source materials. Two categories of source materials were studied based on calcium content and their fineness. For pozzolanic activity, 50% cement was replaced by industrial waste, and mortar cubes were cast with 1:3 proportion and then tested for compressive strength after 7 and 28 days of water curing. Thereafter geopolymer mortar cubes with 1:3 proportions were prepared using various industrial waste and their different combination based on calcium content which was activated by sodium hydroxide and sodium silicate solutions. The water-to-binder ratio of 0.25 to 0.40 was considered based on the fineness of the source material, while the alkaline solution-to-fly ash ratio was maintained constant at 0.35. Sodium hydroxide having 13M concentration and sodium silicate contains 50.38% solid was used as alkaline activators. Geopolymer mortar cubes were cast similar to cement mortar. After demolding, they were kept in an oven at 90°C for 6 hours of curing, and then after a specified period of the rest period, cubes were tested for compressive strength. The test results show that the fineness of the source materials plays important role in the activation process but need more water to binder ratio to maintain the consistency of mix in a plastic state. It is also observed that the calcium content present in source material accelerates the pozzolanic activity of geopolymer mortar and achieved higher compressive strength. Also, the combination of two different source materials based on fineness and calcium content achieved excellent compressive strength.

Keywords: Industrial Waste, Fineness, Pozzolanic Activity, Geopolymer Mortar, Compressive Strength.

[1] *Presenting and corresponding author.*
Research Scholar, Civil Engineering Department, SRES Sanjivani College of Engineering, Savitribai Phule Pune University, Pune 423603, India. *E-mail:* nawalesantosh35@gmail.com
[2] *Corresponding author.*
Professor, Civil Engineering Department, SRES Sanjivani College of Engineering, Savitribai Phule Pune University, Pune 423603, India. *E-mail:* svpatankar11@gmail.com

Effect of Replacing Fine Aggregate with Brick Powder on Mechanical and Durability Property of SCC

Hiwot Tsegaye Wolde[1] and Kizhakkumodom Venkatanaryanan[2]

ABSTRACT

During recent years, the development of self-consolidating concrete (SCC) has caught significant attention in providing sustainable solutions for many in-situ and precast construction applications. The development of SCC containing brick powder as an alternative material by replacing fine aggregate has not been explored much by investigators. In this study, the mechanical and durability of SCC with 5% and 12.5% fine aggregate replacement of brick powder is analyzed. The Compressive strength were measured at the curing age of 7, 28 and 90-day and the rest flexural strength, tensile strength, water absorption, sorptivity, surface resistivity and RCPT tests were measured at the curing age of 28 days.

From the result, best mechanical strength results were obtained from the samples containing 5% and 12.5% brick powder (MB1 and MB2, respectively) as compared to reference specimens (MB0). The brick powder has the highest pozzolanic activity at the curing age of 7, 28 and 90 days. Similarly, a sample with brick powder has shown better resistance towards water absorption, sorptivity, RCPT and also higher surface resistivity value compared to reference mixture. The results indicated that using the brick powder in SCC by replacing fine aggregate does not affect mechanical and durability properties. This can be related with the mineralogical constitute of brick powder and the micro-filling effect of the relatively finer particles of brick powder.

Keywords: Brick Powder, SCC, Sorptivity, RCPT, Mechanical, Durability.

[1] *Presenting & corresponding author.*
Ph.D. Student, Department of Civil Engineering, Kanpur 208016, India.
E-mail: hiwot@iitk.ac.in
[2] *Corresponding author.*
Assistant Professor, Department of Civil Engineering, Kanpur 208016, India.
E-mail: kvharish@iitk.ac.in

Characteristics of Carbapenem Resistant *Acinetobacter Sp.* Strains Isolated from Wastewater and their Possible Transmission into Natural Environment

Jakub Hubeny[1], Monika Harnisz[2] and Ewa Korzeniewska[3]

ABSTRACT

The occurrence and spread of antibiotic resistance and with it multidrug resistant bacteria is an alarming problem and a threat to public health. The emergence of bacterial resistance to antibiotics is common in areas where antibiotics are frequently used. Wastewater and wastewater treatment plants are one of the main reservoirs and spreading sources of antibiotic resistance in the natural environment and a place for the exchange of antibiotic resistance genes between bacteria of different genera or species. *Acinetobacter baumanii* is known as the leading emerging pathogen causing frequent hospital outbreaks. Therefore, the aim of this research was to determine the presence of antibiotic resistance determinants in *Acinetobacter* sp. strains isolated from wastewater at individual stages of the treatment process in three seasons.

The wastewater samples were collected in three periods (February, June, September) from five sampling sites of the wastewater treatment plant treatment process. Using culture methods, 301 strains were obtained, out of 258 strains of the *Acinetobacter* sp. and 21 strains of the species *Acinetobacter baumanii* were identified using molecular methods. Subsequently, with use of molecular tools as qualitative and quantitative PCR reaction and metagenomic analysis, the presence of beta-lactam resistance genes (Ambler class B and D), insertion sequence *ISAba1*, class I, II, and III integron integrase genes and

[1] *Presenting author.*
  M.Sc., Department of Water Protection Engineering and Environmental Microbiology, University of Warmia and Mazury in Olsztyn, Prawocheńskiego 1, Olsztyn 10720, Poland.
  E-mail: jakub.hubeny@uwm.edu.pl

[2] *Corresponding author.*
  Associate Professor, Department of Water Protection Engineering and Environmental Microbiology, University of Warmia and Mazury in Olsztyn, Prawocheńskiego 1, Olsztyn 10720, Poland.
  E-mail: monika.harnisz@uwm.edu.pl

[3] *Corresponding author.*
  Professor, Department of Water Protection Engineering and Environmental Microbiology, University of Warmia and Mazury in Olsztyn, Prawocheńskiego 1, Olsztyn 10720, Poland.
  E-mail: ewa.korzeniewska@uwm.edu.pl

taxonomic structure was clarified The highest detection frequency of *Acinetobacter* sp. and *Acinetobacter baumanii* was determined in June and September. Among the resistance genes, *bla*$_{OXA-51}$ gene revealed the highest incidence in strains of *Acinetobacter* sp. Within 13 strains were found the presence of *ISAba1*/bla$_{OXA-51}$ complex, which confirms the occurrence of carbapenem-resistance strains in analyzed samples. The absence of a reduction of *Acinetobacter* isolates carrying antibiotic resistance genes in treated wastewater compared to influent wastewater suggests that the treatment processes used in the wastewater treatment plant are ineffective. The conducted research confirms that bacteria of the genus *Acinetobacter* can pose a serious epidemiological threat and may spread alarming beta-lactam resistance genes in the environment.

Keywords: *Acinetobacter sp., Antibiotic Resistance, Antibiotic Resistance Genes, Wastewater, Wastewater Treatment Plant.*

Influence of Grading of Aggregates using Particle Packing Methods on Strength and Microstructure of Blended Cement Mortars

Chandra Sekhar Karadumpa[1] and Rathish Kumar Pancharathi[2]

ABSTRACT

In recent times, blended cements having partial replacement of mineral additives like fly ash (FA) and granulated blast furnace slag (GBFS) with Ordinary Portland Cement (OPC) proven to be more sustainable cements with several ecological benefits and increased performance and durability aspects of concrete compared to plain OPC. In this study three blended cements namely Portland Pozzolana Cement (PPC) (OPC + FA), Portland Slag Cement (PSC) (OPC + GBFS) and Composite Cement (CC) (OPC + FA + GBFS) and three Particle Packing Models (PPM) viz. Modified Toufar Model (MTM), JD Dewar model (JDD) and Compressible Packing Model (CPM) are considered to assess the performance of cement mortars. The fine aggregate interaction curves were developed for polydispersed aggregate system using all three PPM methods. The theoretical maximum packing density of fine aggregates obtained through MTM, JDD and CPM models are 0.681, 0.674 and 0.668 respectively. While the experimental packing density values are 0.617, 0.626 and 0.642 respectively. Hence, CPM model is more realistic and best suited for aggregate proportioning as it considers wide range of factors like shape and size of the aggregate, loosening effect, wall effect and compaction effort which is not considered in other particle packing theories. The mortar samples prepared using CPM model, showed an average increase in compressive strength of 20.71% at 7 days, 12.26% at 28 days and 10.29% at 90 days compared to samples designed using IS 650:2008. Similar trend was observed from MTM and JDD mortar samples also with strength of mortars varying as CPM>MTM>JDD>IS 650. The hydrated compound phases and microstructure is examined using the Scanning Electron Microscopy (SEM) and X-Ray Diffraction (XRD).

Keywords: Blended Cements, Particle Packing, Strength, Microstructure.

[1] *Presenting & corresponding author.*
Research Scholar, Department of Civil Engineering, National Institute of Technology Warangal, India.
E-mail: chandukv19@student.nitw.ac.in

[2] *Co-author.*
Professor, Department of Civil Engineering, National Institute of Technology Warangal, India.
E-mail: rateeshp@nitw.ac.in

Integrons Involved in the Co-Selection of Antibiotic Resistance in Heavy Metal Contaminated Environments

Wiktor Zieliński[1] and Ewa Korzeniewska[2]

ABSTRACT

Introduction: Anthropogenic activity greatly contributes to the entry and mixing of heavy metals and antibiotics in the natural environment. As a result, indigenous bacteria are exposed to contact and the impact of these micro-pollutants. Unlike antibiotics, heavy metals are not degraded in the environment and therefore exert long-term selection pressure. Heavy metal contamination in the environment can play an important role in the maintenance and develop of antibiotic resistance, which is known as co-selection. The coexistence of heavy metal and antibiotic resistance genes on the same mobile genetic element such as integrons, transposons, and plasmids is termed co-resistance. The aim of the study was to identify the possibility of the occurrence of co-selection mechanisms of antibiotic resistance in an environment contaminated by heavy metals, as well as to show that the first, second and third class integrons play a fundamental role in the functioning of these phenomenon.

Methods: In this study, the concentrations of antibiotic resistance genes (ARGs), heavy metal resistance genes, integrase-coding genes (*int*I1, *int*I2, *int*I3) as well as concentrations of selected heavy metals in wastewater and river water samples were investigated. The obtained number of genes were used to calculate the correlation using Spearman's rank analysis.

Results: Nickel and cobalt resistance gene – *cnr*A was correlated ($p < 0.05$) with 12 ARGs (*erm*B, *erm*C, *erm*F, *mef*A, *sul*1, *sul*2, *bla*$_{SHV}$, *bla*$_{OXA}$, *tet*A, *aac (6 ')-Ib-cr*, *qep*A, *amp*C) and with genes encoding all three classes of integrase. Moreover, the abundance of each of the ARGs also correlated with at least of one of the integrase genes. Strong statistically

[1] *Presenting author.*
Department of Water Protection Engineering and Environmental Microbiology,
The Faculty of Geoengineering, University of Warmia and Mazury in Olsztyn, Poland.
[2] *Corresponding author.*
Department of Water Protection Engineering and Environmental Microbiology,
The Faculty of Geoengineering, University of Warmia and Mazury in Olsztyn, Poland.
E-mail: ewa.korzeniewska@uwm.edu.pl

significant correlations ($p < 0.05$) were also observed between the abundance of ARGs and five heavy metals – zinc, copper, arsenic, chromium and nickel concentration. Of these, zinc was dominant, correlated with 10 ARGs, and copper correlated with 8 ARGs.

Conclusions: The correlations obtained in this study confirm that the co-selection of antibiotic resistance occurs in the environment in the presence of copper and zinc, as well as heavy metals less concentrated in the environment – arsenic, chromium and nickel. The obtained results also emphasize that integrons play a key role in co-selection mechanisms, which also contribute to the spread of resistance among microorganisms through horizontal gene transfer.

Keywords: Antibiotic Resistance Genes, Integrase Genes, Co-selection, Heavy Metals.

F-EIR Conference 2021 on
Environment Concerns and its Remediation (ECR21)
Chandigarh, India, October 18–22, 2021

Sustainable Design of Textile Industry Effluent Treatment Plant with Constructed Wetland

Aparupa Shenoy[1], Bishnu Kant Shukla[2] and Vaishnavi Bansal[2]

ABSTRACT

Textile has become a major source of income for India's economy. Steps involved in textile manufacturing are fabrication of yarn, fabric production, dying, printing and finishing. Dying, printing and finishing mainly water intensive processes and involves in generation of liquid waste. Major importance of this effluent is due to presence of different salts, acids, alkalis and anthropogenic dyes in good concentration. Many times dyesacts as sources of heavy metals. Effluent from this type of industries not only makes the water body unfit for downstream use but also disturbs the natural aquatic ecosystems. Colored textile effluent after joining any water body make the water unfit for downstream usage as well as interfere with natural activities of aquatic life. Govt. of India is taking strict actions toward reviving major rivers thus becoming more vigilant toward safe disposal standards. Many industries fail to effectively operate their ETP plants due to height installation as well as operation cost. Developing country like India thus not only requires an effective treatment plant but also an economic as well as environment friendly process is also in high demand as a sustainable treatment option.

It has been observed that Constructed Wetlands (CW) provide a economic, easy as well as environment friendly treatment option having the ability to treat event different industrial effluents. CW is 42% economical than activated sludge process where as 80% economical based on operation and maintenance expenses. Here effort has been given to design a sustainable textile industry effluent treatment plant with Constructed Wetland. CW has been designed by various methods, under different flow conditions and indifferent weather conditions. Majorly BOD load is considered as main design parameters in various methods where ask-C* model has the provision of considering the

---

[1] *Presenting author*
Department of Civil Engineering, JSS Academy of Technical Education, Noida, India.
Department of Environmental Engineering, Delhi Technological University, Delhi, India.
E-mail: aparupashenoyy@gmail.com
[2] *Corresponding author*
Department of Civil Engineering, JSS Academy of Technical Education, Noida, India.
E-mail: bishnukantshukla@gmail.com
[3] Department of Civil Engineering, JSS Academy of Technical Education, Noida, India.
E-mail: gvaishnavibansal@gmail.com

COD load for area calculation. The site under study is having a green belt of 1500 sqmts. According to k-C* model while considering the COD load both horizontal as well as vertical subsurface flow CW can be constructed in the green belt. It has also been observed that CW is capable enough to replace the conventional activated sludge process (ASP).

Keywords: Sustainable Design, Textile Industry Effluent, Treatment Plant, CW.

F-EIR Conference 2021 on
Environment Concerns and its Remediation (ECR21)
Chandigarh, India, October 18–22, 2021

Treatability of Effluent from Small Scale Dye Shop Using Water Hyacinth

Aparupa Shenoy[1], Vaishnavi Bansal[2] and Bishnu Kant Shukla[3]

ABSTRACT

Environmentally and economically sustainable water treatment systems are more attractive as many countries and communities are becoming water scares, heavily populated, and decentralized. During last few decades India has become one of the prominent figure in textile production and export. Urbanization and economic stability led to grater inclination toward apparels. India from ancient time being known for its vibrant color and with passage of time the synthetic dyes have taken place of natural dies which is neither environment friendly nor easily degradable. In India many small scale dying shops are there which uses various chemical and synthetic dyes and discharges the effluent directly to municipal waste water system. Though the effluent is less in volume but quite concentrated in nature. Conventional municipal treatment processes are not designed to handle high COD load. This can leads to failure of bio-reactors of the treatment plants. There always lies high chance of damaging the ecosystem of the receiving waterbody, specially isolated like lake or pond, if this type of effluent is disposed for long duration.

Here effort has been given to analyses that weather this type of effluent can be treated in isolation in common pots and by using commonly available plants. The aim of this study is to develop and evaluate the efficiency of lab scale constructed wetland, planted with water hyacinth, to treat dye (azo) rich effluent from small scale textile dye shop. This can be considered as low cost, simple and environment friendly treatment option. The result of this study indicate that the lab scale Constructed Wetland planted with Water Hyacinth can reduce BOD, COD, Turbidity, Conductivity and Dissolved solids to acceptable labels.

Keywords: BOD, COD, Constructed Wetland, Water Hyacinth, Azo Dye.

[1] *Presenting author.*
Department of Environmental Engineering, Delhi Technological University, Delhi, India.
Department of Civil Engineering, JSS Academy of Technical Education, Noida, India.
E-mail: aparupashenoyy@gmail.com
[2] Department of Civil Engineering, JSS Academy of Technical Education, Noida, India.
E-mail: gvaishnavibansal@gmail.com
[3] *Corresponding author.*
Department of Civil Engineering, JSS Academy of Technical Education, Noida, India.
E-mail: bishnukantshukla@gmail.com

Resistance of Geopolymer Concrete against Chemical Attacks

Kumar Lal Babu[1], Prarthita Basu[2] and Ramesh Chandra Gupta[3]

ABSTRACT

Geopolymer concrete (GPC) is new edge concrete and significantly utilised in place of ordinary Portland cement (OPC). Development of GPC lowers the use of OPC. The main advantage of using geopolymer concrete is excellent strength and reduction in CO_2 emission in environment. Present paper evaluated the effectiveness of GPC prepared with fly ash contrary to chemical attacks. In total, 5 geopolymer concrete mixes and one reference mix of M35 grade with ordinary Portland cement were made. In this experimental program, silica fume (SF) was used as an alternate of fly ash (FA). The substitution made in the range of 0%, 5%, 10%, 15% and 20%. The impact of silica fume has been studied by exposing the concrete cubes in 2% H_2SO_4 and 5% NaCl solutions up to 180 days and degradation of concrete was measured with respect to variation in weight and compressive strength. Experimental outcome reveals that loss in compressive strength for reference mix (M35) and GPC (SF20) in 2% H_2SO_4 at 180 days were obtained 40% and 10%. Percentage loss in compressive strength for reference mix (M35) and GPC (SF20) in 5% NaCl at 180 days were found 20% and 3%. Experimental outcomes specified that use of type of binder has a substantial role on resistance of chemical attacks. GPC mix containing 80% fly ash and 20% silica fume develops the best GPC mix. Thus, it may be concluded that the performance of geopolymer concrete with the inclusion of silica fume is enhanced against chemical attacks with respect to reference concrete mix.

Keywords: *Geopolymer Concrete, Fly ash, Silica Fume, Sulfuric Acid, Sodium Chloride, Compressive Strength.*

[1] *Corresponding author.*
Scientist C, Central Soil and Material Research Station, Ministry of Jalsakti, Delhi, India.
E-mail: kumarlalbabu72@gmail.com
[2] *Presenting & corresponding author.*
PhD Scholar, Civil Engineering Department, Malaviya National Institute of Technology, Jaipur, India.
E-mail: sprarthitas@gmail.com
[3] *Corresponding author.*
Professor, Civil Engineering Department, Malaviya National Institute of Technology, Jaipur, India.
E-mail: rcgupta.ce@mnit.ac.inv

Workability and Mechanical Properties of FA-GGBS based Geopolymer Concrete

Kumar Lal Babu[1], Prarthita Basu[2] and Ramesh Chandra Gupta[3]

Abstract

This research work represents an evaluation of an investigational work performed on the workability and mechanical parameters of geopolymer concrete prepared with fly ash (FA) and ground granulated blast-furnace slag (GGBS). This research work consists of partial substitution of FA with GGBS ranging from 0% to 40% at a step increment of 10%. A best suited combination of FA and GGBS was proposed with reference to the results of workability, compressive strength, shear test and pull-out test. The experimental outcomes showed that the inclusion of GGBS improves the viscosity of geopolymer concrete mix which inversely lessen the workability. Replacement of FA with GGBS enhances compressive strength and shear strength. However, in the case of pull-out strength, FA improves the strength but restricted to a lower replacement percentage with GGBS. From experimental investigation, it can be concluded that 20% GGBS may be considered as an ideal substitution with respect to the workability and mechanical strength parameters of geopolymer concrete. The purpose of this work is to apply FA and GGBS in optimum percentage to develop sustainable geopolymer concrete. Production of geopolymer concrete does not involve Portland cement, eventually it will reduce the greenhouse gas emissions with utilization of power plant and stell-iron plants by-products in making of sustainable geopolymer concrete.

Keywords: Geopolymer Concrete, Fly Ash, Ground Granulated Blast-Furnace Slag, Slump Flow, Compressive Strength, Pull-out Strength.

[1] *Corresponding author.*
Scientist C, Central Soil and Material Research Station, Ministry of Jalsakti, Delhi, India.
E-mail: kumarlalbabu72@gmail.com
[2] *Presenting & corresponding author.*
PhD Scholar, Civil Engineering Department, Malaviya National Institute of Technology, Jaipur, India.
E-mail: sprarthitas@gmail.com
[3] *Corresponding author.*
Professor, Civil Engineering Department, Malaviya National Institute of Technology, Jaipur, India.
E-mail: rcgupta.ce@mnit.ac.in

Assessment of Energy Efficiency for Traditional Non-Engineered and Engineered Residential Buildings: A Case of North-Eastern India

Rituparna Pal[1], Souvanic Roy[2] and Biswajit Thakur[3]

ABSTRACT

Huge energy consumption in the building sector and consequent Green House Gas (GHG) emission is a major sustainability concern for India. According to Government of India (GOI) energy statistics, residential and commercial structures consumed 32% of the India's total electricity produced in 2016. The energy demand of India's buildings is predicted to face a hike of 800% in 2047 as compared to 2012. In 2015, India as a party to the United Nations Framework Convention on Climate Change (UNFCCC) held in Paris, voluntarily declared a goal to reduce the intensity of its Gross Domestic Product (GDP) emissions by 20–25%, over 2005 levels, by 2020.

Most of the initiatives taken in the developed countries as well as in India focus on engineered buildings. There is significant gap in the assessment of non-engineered buildings with scant attention for small sized traditional residential buildings. As majority of the buildings (especially Assam-type housing) in the towns and rural areas in the Northeast are primarily constructed with traditional materials and construction techniques, they contribute significantly in the cumulative energy consumption in the country.

The article focuses on non-engineered traditional and engineered, single storey houses of North-Eastern India. These traditional houses, made of Ikra (a locally available reed and effective against earthquake events), wood and mud, are unfortunately becoming extinct with advent of newer technologies. Energy performances of these buildings using traditional and conventional construction materials have been assessed through Whole Building Energy Simulation and the results are compared with iterations in buildings using parametric data prescribed in Eco Niwas Samhita—the Energy Conservation Building Code (ECBC), in India. The study suggests the scope analysed best case building parameters in real life construction, to ensure better efficiency in energy performance and, thereby making the houses less dependent on mechanical/active systems for achieving thermal comfort.

Keywords: *Non-Engineered & Engineered Buildings, North-East India, Eco Niwas Samhita, Whole Building Energy Simulation, Energy Performance Index.*

[1] *Presenting & corresponding author.* PhD Scholar, Indian Institute of Engineering Science and Technology, Shibpur, India. ***E-mail:*** rituparnapaul51@gmail.com
[2] Professor, Indian Institute of Engineering Science and Technology, Shibpur, India. ***E-mail:*** soy1roy@gmail.com
[3] Associate Professor, Meghnad Saha Institute of Technology, India. ***E-mail:*** biswajit.thkr@gmail.com

F-EIR Conference 2021 on
Environment Concerns and its Remediation (ECR21)
Chandigarh, India, October 18–22, 2021

Functionalized PVA Membrane for Separation of Micro-Emulsion

Dharmveer Yadav[1], Arthi Karunanithi[2], Sumit Saxena[3] and Shobha Shukla[4]

ABSTRACT

The release of liquefied hydrocarbons in form of domestic and industrial effluents along with oil spills cause significant adverse effects on the soil, water, aquatic ecosystem and humans. Thus, selective and cost-effective technology to address this challenge is highly desirable. Here, we report fabrication of electrospun polyvinyl alcohol (PVA) membrane, modified with glutaraldehyde (GA) and a device thereof, for treatment of oil-emulsions and recovery of precious fossil fuel. The functionalized PVA membranes are superoleophobic with a high static underwater motor oil contact angle of 163°. Investigation of wetting properties suggest that the membrane can be used for efficient separation of different oils such as sesame oil, motor oil, mustard oil, and sunflower oil from their emulsions. The motor oil emulsion with separation efficiency of >99% at an excellent permeate flux of 5128 L/m^2.h.bar has been achieved. Thus, the prepared PVA membrane construes an easy solution for not only effective treatment of oily wastewater but also for oil recovery with high flux.

Keywords: *PVA Membrane, Oil-water Separation, Oil Pollution, Environmental Remediation, Electrospinning.*

[1] *Presenting author.*

Centre for Research in Nanotechnology and Science, Indian Institute of Technology Bombay, Mumbai 400076, India. *E-mail:* dharmveer.water@iitb.ac.in

[2] Nanostructures Engineering and Modeling Laboratory, Department of Metallurgical Engineering and Materials Science, Indian Institute of Technology Bombay, Mumbai 400076, India.

[3] Nanostructures Engineering and Modeling Laboratory, Department of Metallurgical Engineering and Materials Science, Indian Institute of Technology Bombay, Mumbai 400076, India.

Water Innovation Center: Technology Research & Education, Department of Metallurgical Engineering and Materials Science, Indian Institute of Technology Bombay, Mumbai 400076, India.

[3] *Corresponding author.*

Nanostructures Engineering and Modeling Laboratory, Department of Metallurgical Engineering and Materials Science, Indian Institute of Technology Bombay, Mumbai 400076, India.

Water Innovation Center: Technology Research & Education, Department of Metallurgical Engineering and Materials Science, Indian Institute of Technology Bombay, Mumbai 400076, India.

E-mail: sshukla@iitb.ac.in

Graphical Abstract

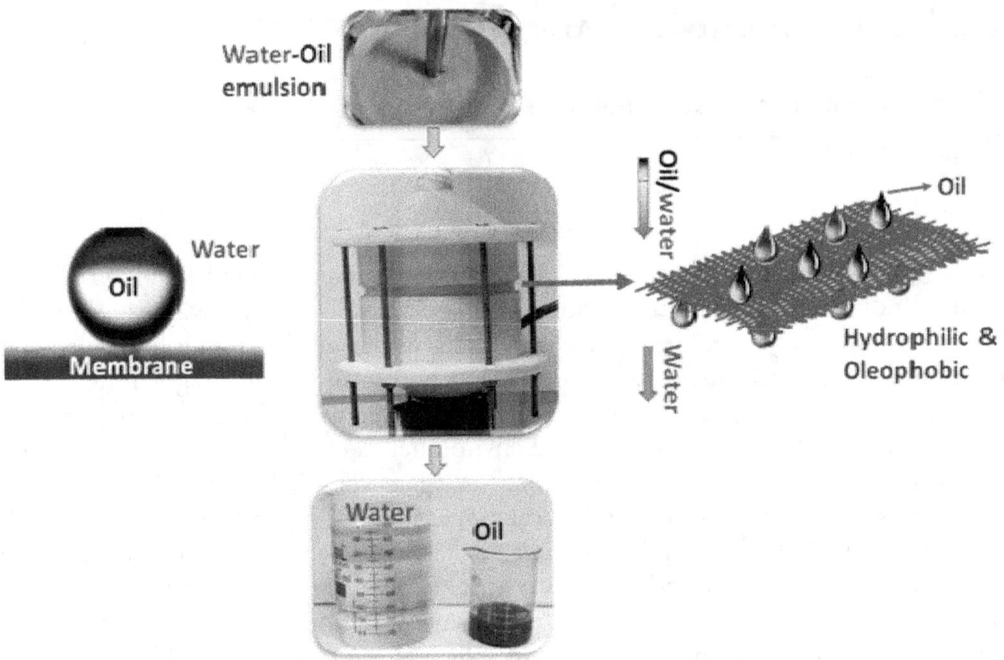

F-EIR Conference 2021 on
Environment Concerns and its Remediation (ECR21)
Chandigarh, India, October 18–22, 2021

Strength and Durability of Dry Sewage Sludge (DSS) as a Replacement for Fine Aggregates

S. Pitchiah Raman[1] and A. Arunkumar[2]

ABSTRACT

Sewage sludge disposal has become a challenging task in India. Central Pollution Control Board of India in 2011 stated that around 10.7 million tons of dry sewage sludge (DSS) is being disposed every year. Generally sewage sludge is disposed by means of incineration, buried in underground pits, disposed into water bodies and used in agricultural lands. These methods lead to air pollution, enhancing global warming through rising atmospheric temperature and affecting the environment's cleanliness. In this study, DSS is introduced in concrete as a replacement for fine aggregates (5, 10, 15 and 20%). All the concrete mixes were tested for its mechanical properties such as compressive strength, split tensile strength, flexural strength, abrasion resistance, impact resistance and durability properties in aggressive conditions like acid attack, and effect of sea water. Results reveal that DSS incorporation of up to 10% has enhanced compressive strength by 9% and replacements till 15% enhanced flexural and split tensile strength up to 13%. In case of durability properties, incorporation of DSS up to 15% showed better results than control concrete in terms of acid attack, effect of sea water, abrasion resistance and impact resistance. Cost analysis indicated that DSS can reduce up to 2.5% of cost considering the efficacy of compressive strength.

Keywords: *Dry Sewage Sludge (DSS), Mechanical and Durability Properties, Aggressive Conditions, Cost Analysis.*

[1] *Presenting & corresponding author.*
Research Scholar, School of Civil Engineering, Vellore Institute of Technology, Chennai, India.
E-mail: s.p.raman273@gmail.com
[2] *Corresponding author.*
Associate Professor, School of Civil Engineering, Vellore Institute of Technology, Chennai, India.
E-mail: a.arunkumar@vit.ac.in

F-EIR Conference 2021 on
Environment Concerns and its Remediation (ECR21)
Chandigarh, India, October 18–22, 2021

A Succinct Review on the Use of NMR Spectroscopy in Monitoring Hydration, Strength Development, and Inspection of Concrete

Rishav Ghosh[1], Mukund Lahoti[2] and Bhoomi Shah[3]

ABSTRACT

NMR is a very useful technique used in analytical chemistry for quality and molecular structure determination of a compound as well as its chemical environment. It works on the principle of the energy gap between two nuclei of an atom for an applied external magnetic field, due to their opposite spins. The potential of the three sub-categories of NMR (H-NMR or proton NMR, liquid-state NMR, and solid-state NMR) for monitoring hydration reaction in cement paste, early-age strength development in fresh concrete, and the inspection of hardened concrete has been elucidated in this paper. Powerful features of NMR in identifying the chemical environment and structure have made it possible for this technique in the evaluation of concrete subjected to sulfate attack and the estimation of the effect of alumino-silicate precursor on the strength of geopolymer. In this manuscript, various other applications of NMR in the non-destructive testing, quality checks, and monitoring of reactions in concrete and cement have been reviewed.

Keywords: Nuclear Magnetic Resonance, Relaxation Times, Cement Hydration, Fresh Concrete, Compressive Strength, Moisture.

[1] *Presenting author.*
Undergraduate Student, Department of Civil Engineering, Birla Institute of Technology and Science, India.
E-mail: f20180487@pilani.bits-pilani.ac.in
[2] *Corresponding author.*
Assistant Professor, Department of Civil Engineering, Birla Institute of Technology and Science, India.
E-mail: mukund.lahoti@pilani.bits-pilani.ac.in
[3] Masters Student, Department of Civil Engineering, Birla Institute of Technology and Science, India.
E-mail: h20180046@pilani.bits-pilani.ac.in

F-EIR Conference 2021 on
Environment Concerns and its Remediation (ECR21)
Chandigarh, India, October 18–22, 2021

Assessing the Monthly Heat Stress Risk to Society Using Thermal Comfort Indices in the Hot Semi-Arid Climate of India

Pardeep Kumar[1] and Amit Sharma[2]

ABSTRACT

Outdoor thermal comfort (OTC) has become a severe issue in the cities as the number of hot days is increasing due to global climate change. The use of urban green infrastructure could be promising in improving the pedestrians' thermal comfort. The aim of the present study is 1) To record the thermal perceptions using subjective evaluation (questionnaire) and onsite monitoring of microclimate parameters in the hot semi-arid (Bsh) climate of Haryana 2) To check the applicability of the Physiological equivalent temperature (PET) and Universal Thermal Climate Index (UTCI) in the Bsh climate of Haryana. 3) To determine the neutral temperature, the neutral temperature range in the Bsh climate of Haryana. 4) To formulate the empirical thermal sensation vote (TSV) model. Biometeorological indices PET and UTCI have been applied to investigate the OTC conditions. The present study used the ASHRAE seven-point sensation scale to record the visitors' thermal sensations. The linear regression analysis has been applied between thermal sensations and applied indices to determine the thermal neutrality of visitors at parks. UTCI range found to be 28.03°C to 35.61°C while PET range found to be 24.04°C to 37.55°C. The neutral PET and UTCI was found to be 30.79°C and 31.82°C, respectively. By using correlations, it was observed that the UTCI is a better predictor of comfort conditions in the present study. Empirical TSV model predicted that the dry bulb temperature as the most significant parameter affecting thermal comfort followed by mean radiant temperature. The finding of the present study could be helpful in theoretical design reference in designing the comfortable outdoor public places.

---

[1] *Presenting & corresponding author.*

Doctoral Candidate, Department of Mechanical Engineering, Deenbandhu Chhotu Ram University of Science and Technology, Sonipat, India.

E-mail: kumar.pardeepmech@gmail.com

[2] *Corresponding author.*

Assistant Professor, Department of Mechanical Engineering, Deenbandhu Chhotu Ram University of Science and Technology, Sonipat, India.

E-mail: amitsharma.me@dcrustm.org

Extreme weather conditions, especially heatwave, are a threat to society, affecting livability, wellbeing, and social interactions. The present study aims to assess the monthly heat stress in the outdoor environment from 2010 to 2019 in Sonepat's municipality, representing a hot semi-arid climate. The authors applied three heat stress indices, namely, Wet bulb globe temperature (WBGT), Physiological equivalent temperature (PET), and Universal thermal climate index (UTCI), to estimate the grade of heat stress. While calculations, the highest average WBGT was found in July (33.4 ± 0.77°C), demonstrating July in the "Extreme heat stress" category. The highest mean PET was found in June (42.47 ± 2.34°C), indicating June in the "Extreme heat stress" category. The highest mean UTCI was found in June (38.58 ± 1.82°C), demonstrating "Very strong heat stress." The dry bulb temperature was found to be the most dominant parameter among meteorological parameters promoting extreme heat stress. It was concluded that extreme heat stress was observed in the Pre-monsoon hot weather season and summer monsoon season (especially in June), making the population vulnerable to mortality and morbidity. The findings could provide valuable information to people from various disciplines like Climate scientists, landscape designers, architects, and all relevant stakeholders to develop a heatwave action plan against adverse heat stress.

Keywords: *Outdoor Thermal Comfort, Urban Parks, PET, UTCI, Neutral Temperature Range.*

F-EIR Conference 2021 on
Environment Concerns and its Remediation (ECR21)
Chandigarh, India, October 18–22, 2021

Role of Environment on the User's Behavior in a Campus Public Space

K. Swetha[1] and A. Meenatchi Sundaram[2]

ABSTRACT

Campus life plays a major role in shaping an individual, who plays a major role in influencing our future. The Environment provided for an Individual should be imbibed with the traits and practices of social sustainability in order to ensure a safe and progressive future. Social sustainability comprehends to Human activities with respect to its setting inclusive of built and unbuilt environment. A better understanding of human-environment relationship would give us a clear idea of the needs which has to be catered without compromising the well-being of the other. The main aim of the study is to understand the User behavior in a Setting by applying Integral Theory. The methodology applied is Observer-Based Environment Assessments (OBEA) used for measuring the Quality of the Setting. The setting selected is the pathways of NIT Trichy Campus, the Use behavior has been mapped, User perception being gathered through Survey, and activities in the setting are distinguished by Integral Theory. The study predicts three prominent observed behavior in the particular setting, the survey analyzed the Quality of the setting with respect to User perception and suggested that the quality of the setting and the activity in the pathways are directly proportional, Integral Theory highlights the dimensions that ensures activity in the particular setting. The paper helps us to understand that catering to the needs of the User will develop a sense of ownership towards the setting and positively influencing the environment.

Keywords: Observer Based Environmental Assessment, Setting, Social Sustainability, User Perception, Integral Theory, Pathways.

[1] *Corresponding & presenting author.*
Research Scholar, Department of Architecture, NIT Trichy, Tiruchirappalli, India.
E-mail: shwethasrm@gmail.com
[2] Associate Professor, Department of Architecture, NIT Trichy, Tiruchirappalli, India.
E-mail: meenatchi@nitt.edu

F-EIR Conference 2021 on
Environment Concerns and its Remediation (ECR21)
Chandigarh, India, October 18–22, 2021

Effects of Soil Type on Contaminant Transport in the Aquifer System: A Numerical Investigation Using 2D Mobile-Immobile Model

Abhay Guleria[1] and Sumedha Chakma[2]

ABSTRACT

We have investigated 2-D nonpoint source contaminant transport behaviour in the aquifer system (100 m × 50 m) using a mobile-immobile (MIM) model with variable dispersion function for various soil-type (sand, silt, clayey loam, and sandy loam). The temporal evolution of concentration profiles and breakthrough curves were compared for conservative and reactive cases due to pulse-type and continuous source boundary conditions. Zeroth and first temporal moments were computed to investigate the effect of soil-type on plume evolution dynamics. A significant variation in the magnitude and spatio-temporal distribution of contaminant plume was observed between low (silt, clayey loam, sandy loam) and high (sand) hydraulic conductivity soil-type. The spreading of a contaminant plume in the transverse direction was dominant in the clayey loam and sandy loam compared to sand soil-type. The maximum value of a zeroth temporal moment of reactive contaminant concentrations was found to be much lower (2–8 order less) than the conservative case and followed the order as sand > silt > clayey loam soil-type. The first temporal moment of reactive contaminant for all soil-type was found to be higher than the conservative case. The dual-porosity model-based approach implemented in this study can be used for scenarios where fluctuations in the water table occur.

Keywords: Soil Type, Stagnant Region, Reactive Contaminant, Dual-Porosity Model, Contaminant Mass Recovery.

[1] *Presenting & corresponding author.*
Department of Civil Engineering, Indian Institute of Technology, Delhi 110016, India.
E-mail: abhayguleria92@gmail.com; abhay_guleria@civil.iitd.ac.in
[2] Department of Civil Engineering, Indian Institute of Technology, Delhi 110016, India.

Synthesizing a Leak Proof and Stable MOF via Silica Layer Addition

Saif Ul Mehdi[1] and Kannan Aravamundan[1]

ABSTRACT

Metal Organic Frameworks (MOFs) are a new class of materials. These were recently developed adsorbents which have wide area of application such as – catalysis [1], sensing [2], waste water treatment [3], drug delivery [4], etc. There are two ways of synthesizing them – conventional and alternative. The conventional process of synthesis involves application of electrical heat treatment. They are known as hydrothermal/ solvothermal depending on whether the solvent being used is water or other organic solvents like ethanol, dimethyl formamide, diethyl formamide, etc. The alternate process of synthesis involves application of electric potential, ultrasound, microwaves or mechanical force. The conventional process is used to obtain high purity MOFs involving single crystal while the alternate process is used to obtain microcrystalline powder. For few MOFs, the amount of energy required to form the final product is very high indicating high activation energy which would make it difficult to obtain the MOF [5]. For such MOFs alternate methods were developed. The alternate methods also allow to obtain nano-sized crystallites for the MOFs which were obtained via conventional process. The advantage of MOFs over other porous materials as adsorbent is that their pores are tunable and the functionalization to improve the selectivity for a particular pollutant is easy. Some of the MOFs are found to be not stable in the acidic or basic medium [6] and leaching of the metal center is prevalent in different MOFs [7, 8].

In this work, we have synthesized TMU-5 using an optimized sonochemical process and then coated with silica layer to reduce the leakage of metal center. The virgin MOF and silica coated MOF were characterized using XRD, FTIR, SEM, Point of Zero Charge (PZC) and it was found that there was reduction in point of zero charge from 6.96 for TMU-5 to 5.26 for SiO_2@TMU-5 for pH 3. The performance of these adsorbents were checked by adsorbing cadmium and nickel on them. It was found that the adsorption performance was affected and the equilibrium loading reduced from 1929 mg/g to 634 mg/g for cadmium.

[1] *Presenting & corresponding author.*
Department of Chemical Engineering, Indian Institute of Technology Madras, Chennai, India.
E-mail: ch16d200@smail.iitm.ac.in
[2] Department of Chemical Engineering, Indian Institute of Technology Madras, Chennai, India.

[1] Drout, R.J., Robison, L. and Farha, O.K., Catalytic applications of enzymes encapsulated in metal–organic frameworks, *Coord. Chem. Rev.*, 381 (2019), 151–160. https://doi.org/10.1016/J.CCR.2018.11.009.

[2] Zhu, S.Y. and Yan, B., A novel sensitive fluorescent probe of S2O82- and Fe^{3+} based on covalent post-functionalization of a zirconium(iv) metal-organic framework, Dalt. Trans. 47 (2018) 11586–11592. https://doi.org/10.1039/c8dt02051e.

[3] Canivet, J., Fateeva, A., Guo, Y., Coasne, B., Farrusseng, D., Water adsorption in MOFs: Fundamentals and applications, *Chem. Soc. Rev.*, 43 (2014) 5594–5617. https://doi.org/10.1039/c4cs00078a.

[4] Lakshmi, B.A. and Kim, S., Current and emerging applications of nanostructured metal–organic frameworks in cancer-targeted theranostics, Mater. *Sci. Eng. C.* 105 (2019) 110091. https://doi.org/10.1016/J.MSEC.2019.110091.

[5] Stock, N., Biswas, S., Synthesis of metal-organic frameworks (MOFs): Routes to various MOF topologies, morphologies, and composites, *Chem. Rev.*, 112 (2012) 933–969. https://doi.org/10.1021/cr200304e.

[6] Ma, M., Lu, L., Li, H., Xiong, Y. and Dong, F., Functional metal organic framework/ SiO_2 nanocomposites: From versatile synthesis to advanced applications, Polymers (Basel). 11 (2019). https://doi.org/10.3390/polym11111823.

[7] Ke, F., Qiu, L.G., Yuan, Y.P., Peng, F.M., Jiang, X., Xie, A.J., Shen, Y.H., Zhu, J.F., Thiol-functionalization of metal-organic framework by a facile coordination-based postsynthetic strategy and enhanced removal of Hg 2+ from water, *J. Hazard. Mater.* 196 (2011) 36–43. https://doi.org/10.1016/j.jhazmat.2011.08.069.

[8] Zhang, J., Xiong, Z., Li, C. and Wu, C., Exploring a thiol-functionalized MOF for elimination of lead and cadmium from aqueous solution, *J. Mol. Liq.*, 221 (2016), 43–50. https://doi.org/10.1016/j.molliq.2016.05.054.

Keywords: *Metal Organic Framework, Adsorption, Silica Layer Addition, Leaching, Sonochemical.*

F-EIR Conference 2021 on
Environment Concerns and its Remediation (ECR21)
Chandigarh, India, October 18–22, 2021

Durability of Rice Husk Ash and Basalt Fibre based Sustainable Geopolymer Concrete in Rigid Pavements

Mahapara[1] and Gyanendra Singh[2]

ABSTRACT

In this analysis, the Rice husk ash (RHA) was used as source material in geopolymer concrete in alkaline activator presence formed by blending of sodium hydroxide and sodium silicate. Basalt fibres (B. F'S) were used as fibre reinforcement. Rice husk ash is silica luxurious, and basalt fibres are alumina luxurious, and this combination proved better for rigid pavement durability. The sodium silicate (SST) to sodium hydroxide (SHD) ratio and alkaline activator to binder content ratio was kept as 0.4 and 2.0, respectively, the concentration of sodium hydroxide was kept as 14M, Superplasticizer (S.Z.) was used to enhance the fluidity of geopolymer concrete, and the source material used was 500 kg/m^3, in mix RHAM. The different percentages of B. F'S used were 1, 5,10,15,20 and 25% in combinations RHB1, RHB5, RHB10, RHB15, RHB20, and RHB25 and the results of mixes with B.F'S were compared with the control mix OPCM, which was prepared as for ordinary Portland cement concrete of grade 40 with water to cement ratio of 0.35. The results revealed that the durability characteristics like water absorption, acid attack, chloride attack and sulphate attack were less in mix RHB10 as compared to the control blend and all other blends. RHAM also gave better results for durability characteristics as compared to control mix OPCM.

Keywords: *Water Absorption, Acid Attack, Chloride Attack, Sulphate Attack.*

[1] *Presenting & corresponding author.*
Department of Civil Engineering, DCRUST, Sonipat 131039, Haryana, India.
E-mail: mahaparaabbas@gmail.com
[2] *Corresponding author.*
Department of Civil Engineering, DCRUST, Sonipat 131039, Haryana, India.
E-mail: singhgyan27@yahoo.in

F-EIR Conference 2021 on
Environment Concerns and its Remediation (ECR21)
Chandigarh, India, October 18–22, 2021

Impact Strength of Rice Husk Ash and Basalt Fibre based Sustainable Geopolymer Concrete in Rigid Pavements

Mahapara[1] and Gyanendra Singh[2]

ABSTRACT

In this analysis, the alkaline activators prepared by mixing sodium hydroxide and sodium silicate and rice husk ash (RHA) were used as the source material in geopolymer concrete. As fibre reinforcement in geopolymer concrete, basalt fibres (B.F.s) were used. Silica luxurious RHA and alumina luxurious B.F.s proved the best mishmash for rigid pavement abrasion resistance. The ratio of sodium silicate (sdsi) to sodium hydroxide (sdhd), alkaline activator to binder content was maintained as 2.0 and 0.4 respectively, the concentration of sodium hydroxide was maintained as 14 M, the fluidity of geopolymer concrete was improved by superplasticizer, and RHA used was 500 kg/m^3 in MRM (control mix which is rice husk ash based geopolymer concrete without basalt fibres). B.F.s were used as 1, 5,10,15,20, and 25 per cent by weight of RHA in the MB1, MB5, MB10, MB15, MB20 and MB25 combinations, and the results of B.F.'s blends were compared with the MRM. The results showed that, relative to the control mix and all other mixes, the impact strength in mix MB10 was high. In contrast to the MRM control combination, the seven, twenty-eight, and ninety-day impact strength of MB10 showed a substantial increase of more than 88, 89, and 92 per cent, respectively.

Keywords: Rice Husk Ash, Basalt Fibre, Abrasion Resistance, Geopolymer Concrete.

[1] *Presenting & corresponding author.*
Department of Civil Engineering, DCRUST, Sonipat 131039, India. *E-mail:* mahparaabbas@gmail.com
[2] *Corresponding author.*
Department of Civil Engineering, DCRUST, Sonipat 131039, India. *E-mail:* singhgyan27@yahoo.in

Seasonal Variation and Source Identification of PM10 in an Industrialised City

Pallavi Pradeep Khobragade[1] and Ajay Vikram Ahirwar[2]

ABSTRACT

A one year PM_{10} sampling campaign from January 2019 to December 2019 is conducted in an urban industrial area Raipur, India. For the development of effective pollution control strategies, quantitative assessment of potentially toxic metals and accurate identification of the pollutant sources are essential. The annual average PM_{10} aerosol mass concentration during the study period is found to be 187.89 ± 91 µg/m³ PM10 concentrations were found to vary with meteorological parameters such as wind speed, temperature, rainfall and humidity. Heavy metals including Fe, Mn, Ni, Pb, Zn, Al, Ba, Cr, Sr, Ga, Mg and V are studied using atomic absorption spectroscopy. Seasonally, average winter time PM_{10} concentration (234.86 ± 104.3 µg/m³) is found higher as compared to summer (169.02 ± 27.1 µg/m³) and rainy season (106.28 ± 32.6 µg/m³) during the study period. The aerosol sources identified by positive matrix factorization (PMF) model five-factor solution are: Iron & steel production (34.7%), vehicular emissions (29.3%), road dust (14.6%), heavy oil/petroleum (14.6%) and coal combustion (6.9%). Among these sources, Iron & steel production contributed maximum in PM_{10} concentrations. The probable source locations were recognized through concentration-weighted trajectory (CWT) analysis. The AirQ+ model was used to assess the health effects of regional PM10 concentrations.

Keywords: PM_{10}, Source Apportionment, Heavy Metals, Back-Trajectory Analysis, HYSPLIT Model.

[1] *Presenting & corresponding author.*
Research Scholar, Civil Engineering Department, National Institute of Technology, Raipur, India.
E-mail: khobragadepallavi19@gmail.com
[2] Assistant Professor, Civil Engineering Department, National Institute of Technology, Raipur, India.
E-mail: avahirwar.ce@nitrr.ac.in

F-EIR Conference 2021 on
Environment Concerns and its Remediation (ECR21)
Chandigarh, India, October 18–22, 2021

Biodegradation of Lac – Soil Microbiome Approach

Thamilarasi Kandasamy[1], Mohammad Fahim Ansari[2],
Sajiya Ekbal[3] and Kewal Krishan Sharma[4]

ABSTRACT

Lac resin is the only natural resin of animal origin exclusively from the lac insect genus, *Kerria*. Sticklac comprising of lac insects and host plant material is processed to produce seedlac, a semi refined form and further purified to shellac, button lac and bleached lac etc. All these forms of lac are widely utilized in numerous fields such as food, pharmaceuticals, cosmetics, varnishing and electrical industries. Although synthetic alternatives are available in the market, natural lac resin has its fixed share in the market due to its non-toxic and biodegradable nature. To understand its biodegradation rate and mode of biodegradation, soil burial method of seedlac was followed. Soil microbes are the eventual key players in biodegradation of any material on the earth. To explore the actual microbial players involved in biodegradation of seedlac, soil microbiome of seedlac buried sample and control sample was documented. V3-V4 Variable region of bacterial barcode marker gene, 16S rRNA of the samples was amplified, sequenced using illumina Hiseq 2500 NGS platform and analyzed. Although obvious variations were not obtained between the samples, apparently the count of Proteobacteria and Acidobacteria were slightly more in seedlac buried soil compared to the control soil. Among Proteobacteria, the count of Betaproteobacteriales is slightly higher in seedlac buried soil samples compared to control, pointing out that the Betaproteobacteriales, would have key role in lac biodegradation.

Keywords: *Seedlac, Shellac, Microbiome, Proteobacteria, Acidobacteria, Lac Insect.*

---

[1] *Presenting & corresponding author.*
Lac Production Division, ICAR-Indian Institute of Natural Resins and Gums, Ranchi, India.
E-mail: k.thamilarasi@gmail.com
[2] Processing and Product Development Division, ICAR-Indian Institute of Natural Resins and Gums, Ranchi, India. *E-mail:* mfansari@rediffmail.com
[3] Lac Production Division, ICAR-Indian Institute of Natural Resins and Gums, Ranchi, India.
E-mail: sajiya.chand@gmail.com
[4] The Director, ICAR-Indian Institute of Natural Resins and Gums, Ranchi, India.
E-mail: kewalkks@gmail.com

F-EIR Conference 2021 on
Environment Concerns and its Remediation (ECR21)
Chandigarh, India, October 18–22, 2021

Problems and Prospects of Heritage based City Development in India

Shipra Goswami[1] and Ashwani Kumar[2]

ABSTRACT

India has been widely recognized for its rich and diverse heritage. Because of the same the country has been a destination spot for many international tourists as well. The heterogeneity of urban fabric specially in heritage zones is a magnificent paradigm to analyse it on all parameters based on pedagogy of imperial practice. The concept of heritage in Indian Cities is challenged because of several factors such as rapid urbanization, increase in the demand for a household in the cities, cultural changes, climatic changes, and various other factors that emphasize pressure on the cities.

India's GDP include 7.3 percent of the tourism industry that accounts for 6.5 percent of total export. This sector also generates 2.7 percent of the total employment in the economy. Since tourism is directly linked with the heritage can help in generating more economy and employment for the people living in the city. However, heritage relies on the tourism directly or indirectly for its livelihood. But the city historic places are unable to balance the development between the old and new, that is harnessing the interest of tourists. The magnanimous manifestation can only be practiced by the people living in the core area. The solution to such challenge for generating more and more economy and employment can be generated by integrating tourism with heritage and making city responsive to development.

This paper thus intends to understand the pressure of development on heritage fabric by identifying the issues and challenges that has been faced by the heritage precincts due to increasing urbanization, tourism. However, the focus of this paper is towards developing a framework and linking the missing link of the various governmental institutions that can help in generating the economy for the people at the same time can achieve balance between heritage conservation and tourism integrated city development.

Keywords: City Development, Heritage, Tourism, Urbanization.

[1] *Presenting & corresponding author.*
Masters in Urban Planning, Department of Architecture & Planning, MNIT Jaipur, India.
E-mail: 2019PAR5564@mnit.ac.in
[2] *Corresponding author.*
Assistant Professor, Department of Architecture & Planning, MNIT Jaipur, India.
E-mail: akumar.arch@mnit.ac.in

Statistical Analysis of Sustainable Geopolymer Concrete

Y.S.N. Kishore[1*], Sai Geeta Devi Nadimpalli[1#], Ashish Kumar Potnuru[1@],
Jayaprakash Vemuri[2] and Mohd Ataullah Khan[3]

ABSTRACT

Geopolymer concrete is an environment-friendly material and is presently accepted as an alternative to conventional concrete. It utilizes industrial by-products like fly ash and slag to reduce CO_2 emissions associated with cement production. Due to the variability of fly ash composition and concrete casting conditions, significant variability is observed in the performance of the concrete produced with such materials. In this study, a non-linear regression technique is used to develop predictive models for the compressive strength of high and low calcium fly ash geopolymer concrete. The criteria such as root mean squared error (RMSE), the statistical significance of model coefficients, and the R-squared values were adopted. The developed models were validated using plots comparing the experimental and predicted response and plots showing residuals. The dependencies of the significant parameters on the compressive strength were highlighted using contour and adjusted response plots. From the contour plots, it was observed that, for geopolymer concrete sourced from the high calcium fly ash, compressive strength was found to increase with the molarity of sodium hydroxide, the activator-to-binder ratio, and the curing temperature; while it decreases with the increase in coarse aggregate content. Similar contour plots, for the case of geopolymer concrete sourced from low calcium fly ash, exhibited improvement in compressive strength with the molarity of sodium hydroxide, the activator-to-binder ratio, and the ratio of Na_2SiO_3 – NaOH solutions; at the same time, it decreases with the increase in water-to-geopolymer solids ratio. For low calcium fly ash geopolymer concrete, the proposed model predicts that compressive strength of 90 MPa could be obtained for fly ash with oxide composition comprising a silica-to-alumina ratio of 4. The regression models developed in the present study will be helpful to develop a design mix for geopolymer concrete comprising either high calcium fly ash or low calcium fly ash.

Keywords: Regression Analysis, Alkali-activation, Geopolymer Concrete, Compressive Strength, Class F Fly Ash, Class C Fly Ash.

[1] Undergraduate Student, Department of Civil Engineering, Mahindra University, Ecole Centrale College of Engineering, Hyderabad, India.
E-mail: [*]kishore.sai248@gmail.com; [#]saigeetadevi160131@mechyd.ac.in; [@]ashish160133@mechyd.ac.in
[2] Assistant Professor, Department of Civil Engineering, Mahindra University, Ecole Centrale College of Engineering, Hyderabad, India. *E-mail:* jayaprakash.vemuri@mahindrauniversity.edu.in
[3] *Presenting & corresponding author.*
Assistant Professor, Department of Civil Engineering, Mahindra University, Ecole Centrale College of Engineering, Hyderabad, India. *E-mail:* mohdataullah.khan@mahindrauniversity.edu.in

F-EIR Conference 2021 on
Environment Concerns and its Remediation (ECR21)
Chandigarh, India, October 18–22, 2021

Evaluating the Combined Effect of Recycled Aggregate and Rice Husk Ash on Concrete Properties

Monalisa Behera[1] and Md. Reyazur Rahman[2]

ABSTRACT

This paper evaluates the synergistic influence of utilizing agricultural wastes i.e. rice husk ash (RHA) and coarse recycled concrete aggregates (CRCA) on the properties of recycled aggregate concrete (RAC). The RAC mixes were made with 0%, and 100% substitution of coarse natural aggregates by CRCA. Cement is further substituted by RHA at 0%, 5%, 10%, 15%, 20%, 25% and 30% to understand the influence of RHA on the fresh and hardened properties of RAC. The mechanical properties were investigated through compression, tension and flexure test at various ages such as 7, 28 and 56 days and other hardened properties like the water penetration resistance and the dry density were also studied. The various properties of the RAC were found to be improved as the RHA quantity is raised till 25% and further decreased when RHA content is increased to 30%. The use of RHA has proved to be very effective as the compressive strength improved by about 26% at 25% RHA. The correlation among the different strength parameters in the RAC mixes has also been established. The water permeability of the concrete mixes got reduced and the dry density increased with the increase in RHA content due to the reduced porosity. This was possible because of the pozzolanic reaction of amorphous silica present in RHA. Hence, the short comings of RAC can be overcome with the use of low end pozzolanic materials like RHA and can be used for retaining the performance of concrete containing 100% RCA.

Keywords: *Recycled Concrete Aggregate, Rice Husk Ash, Sustainability, Mechanical Properties, Permeability, Pozzolanic Material.*

[1] *Presenting & corresponding author.*
CSIR-Central Building Research Institute, Roorkee 247667, India.
E-mail: monalisabehera7@gmail.com
[2] CSIR-Central Building Research Institute, Roorkee 247667, India.
E-mail: engg.reyaz@gmail.com

F-EIR Conference 2021 on
Environment Concerns and its Remediation (ECR21)
Chandigarh, India, October 18–22, 2021

Removal of Ammonium by Products from the Effluent of Bio-cementation System through Struvite Precipitation

Sivakumar Gowthaman[1], Mohsenzadeh Arash[2],
Kazunori Nakashima[3] and Satoru Kawasaki[3]

ABSTRACT

Microbial induced carbonate precipitation (MICP) (also referred to as bio-cementation) is an innovative soil improvement technology, introduced through Bio-mediated Geotechnics. During the process, supplied microorganisms hydrolyze urea, resulting in the formation of calcium carbonate. The precipitates tend to cement the soil particles at particle contacts and enhance the strength of the matrix. Despite the considerable interests in MICP, one major hurdle that hinders the field-scale applications is related to the production of ammonium during the process. The objective of this study is to evaluate the feasibility of struvite ($NH_4MgPO_4 \cdot 6H_2O$) precipitation to eliminate the ammonium from the reaction effluent system. For this purpose, the study is considered in two stages: (i) rinsing the ammonium from the sand, and (ii) precipitating ammonium as struvite. In the first-stage, the conditions of rinsing are studied to optimize the removal of ammonium from the soil. In the second-stage, influences of pH conditions, molar ratio and calcium ions on the precipitation of struvite are evaluated. The study demonstrates, through the struvite precipitation technique, around 90% of ammonium could be removed from the effluent. Finding also suggests that the molar ratio (NH_4^+: Mg^{2+}: PO_4^{3-}) of 1: 1.2: 1 provides a desirable environment for appropriate removal.

Keywords: Microbial Induced Carbonate Precipitation, Soil Improvement, Ammonium Removal, Struvite, Precipitation, Sustainability.

[1] *Presenting & corresponding author.*
Department of Engineering Technology, University of Jaffna, Kilinochchi 44000, Sri Lanka.
E-mail: gowtham1012@outlook.com
[2] Department of Civil and Environmental Engineering, Amirkabir University of Technology, Tehran 159163-4311, Iran.
[3] Division of Sustainable Resources Engineering, Faculty of Engineering, Hokkaido University, Sapporo 060-8628, Japan.

Performance Evaluation of Self-Compacting Concrete Comprising Ceramic Waste Powder as Fine Aggregate

Lilesh Gautam[1], Nitin Meena[2], Jinendra Kumar Jain[3],
Pawan Kalla[4] and Thamer Alomayri[5]

ABSTRACT

As the population around the world is on the rise so to meet consumer's needs a large amount of solid waste is generating and it needs large landfill sites for disposal of this waste. Alteration of this solid waste into other alternative resources will reduce the load on non-renewable resource materials and this can help in solving the landfills problem. Numerous studies have drawn attention to the usage of solid waste in concrete production, which is an imperative material for construction. Specifically, those that can replace fine aggregate because these days many states have banned the use of river sand so researchers all over the globe looking for a suitable material that can replace river sand and also provide sustainability with better strength results. In this study, bone china cup plate-type ceramic waste powder was utilized as a fine aggregate. Ceramic waste powder (CWP) creates environmental degradation and the use of CWP in concrete as a substitute for natural resources would have a positive result on the environment. In this study, the sand was replaced partially with CWP at replacement levels of 0%, 10%, 20%, 30%, 40%, and 50% by the weighted amount of sand. To evaluate various properties of SCC mixes with CWP, various workability tests (i.e. slump flow, v-funnel, j-ring, and l-box) were performed in a fresh state of concrete, while compressive strength, ultrasonic pulse velocity, flexural strength, and water absorption tests were computed to measure the hardened characteristics of SCC mixes. In SCC mixes, the CWP confederation showed an improvement in mechanical strength of concrete at the 20% CWP placement stages and a decline above 20% CWP replacements in all strength characteristics.

Keywords: Fine Aggregate, Ceramic Waste Powder, Self-compacting Concrete, Sustainable Material, Fresh Properties, Mechanical Properties.

[1] *Presenting author.* Research Scholar, Department of Civil Engineering, Malaviya National Institute of Technology, Jaipur 302017, India. *E-mail:* 2017rce9028@mnit.ac.in

[2] M.Tech Student, Department of Civil Engineering, Malaviya National Institute of Technology, Jaipur 302017, India. *E-mail:* 2019pct5125@mnit.ac.in

[3] Associate Professor, Department of Civil Engineering, Malaviya National Institute of Technology, Jaipur 302017, India. *E-mail:* 2019pct5177@mnit.ac.in

[4] M.Tech Student, Department of Civil Engineering, Malaviya National Institute of Technology, Jaipur 302017, India. *E-mail:* jkjain.ce@mnit.ac.in

[5] Department of Physics, Faculty of Applied Science, Umm Al-Qura University, Makkah 21955, Saudi Arabia. *E-mail:* tsomayri@uqu.edu.sa

F-EIR Conference 2021 on
Environment Concerns and its Remediation (ECR21)
Chandigarh, India, October 18–22, 2021

Analysis of Spatial and Temporal Variability of Rainfall Using GIS in Rasipuram Taluk, Tamil Nadu, India

P. Mageshkumar[1], K. Angu Senthil[2], N. Sudharsan[3],
K. Poongodi[4] and P. Murthi[5]

ABSTRACT

In this paper, the variation is spatial and temporal scales of rainfall in and around Rasipuram Taluk, Tamil Nadu, India is presented in which agriculture is a major land use category. Daily rainfall data are collected for seven rain gauge stations namely, Kullampatti, Salem Junction, Sankagiri, Rasipuram, Senthamangalam, Thiruchengode and Gangavalli. The rainfall data are sorted into monthly, seasonal and average annual rainfall categories for assessment. The spatial distribution of rainfall during all the major seasons and average annual rainfall are plotted for better understanding of the spatial variability. Monthly rainfall analysis demonstrates that October month receives highest rainfall whereas lowest rainfall would usually be registered during January at all seven rain gauge stations considered. Annually, the highest rainfall is received in 2015 and least amount of rainfall is received in 2006. Northeast monsoon has the highest average rainfall of 374 mm when compared with all the other seasons during the study period from 2006 to 2015. Spatial analysis reveals that the major portion of the Rasipuram Taluk belongs to moderate rainfall category except in north east monsoon season. The regional level term analysis of rainfall variation can depict exact trend which can be used for agricultural development plans in semiarid regions.

Keywords: Rainfall Variation, Spatial and Temporal Analysis, Rasipuram Taluk, GIS.

[1] Department of Civil Engineering, K.S. Rangasamy College of Technology, Namakkal, India.

[2] Department of Civil Engineering, K.S. Rangasamy College of Technology, Namakkal, India.

[3] Department of Civil Engineering, Vidya Jyothi Institute of Technology, Hyderabad, India.

[4] *Corresponding author.*
Department of Civil Engineering, SR University, Warangal, India.
E-mail: k.poongodi@sru.edu.in

[5] *Presenting author.*
Department of Civil Engineering, SR University, Warangal, India.
Centre for Construction Methods and Materials, SR University, Warangal, India.

F-EIR Conference 2021 on
Environment Concerns and its Remediation (ECR21)
Chandigarh, India, October 18–22, 2021

Evaluation of Barrier for Promoting Green Building Technologies in Coimbatore as Smart City

S. Hema[1], K. Poongodi[2] and P. Murthi[3]

ABSTRACT

Developing and promoting smart cities in India is the dream project of Government of India. Coimbatore is one among the city identified as smart city project for improving the existing infrastructure and constructing green building in order to improve the habitation and energy efficiency through utilization of natural resources. Implementing the green building and energy efficient building concepts in construction industry are the major focus with the aid of technological development. The transformation from conventional to the futuristic approach is the major task in the construction sector. This sector alone requires 40% of natural raw materials, 25% of water and 35% energy resources, as well as emitting 40% of wastes and 35% of greenhouse gases. Considering these parameters, a perception survey was conducted through a descriptive questionnaire among a set of professionals in the construction industry in Coimbatore. The questionnaire was framed with closed ended questions using the Likert Scale. The paper was intended to furnish the detailed investigation to identify the various barriers and suggest the measures to be taken to promote the green building technologies for a sustainable development in smart city mission.

Keywords: *Green Building, Smart City, Barriers in Construction Sector, Sustainable Development, Likert Scale.*

[1] *Corresponding author.*
Department of Civil Engineering, Sri Ramakrishna Engineering College, Coimbatore, India.
[2] Department of Civil Engineering, SR University, Warangal, India.
E-mail: k.poongodi@sru.edu.in
[3] *Presenting author.*
Department of Civil Engineering, SR University, Warangal, India.
Centre for Construction Methods and Materials, SR University, Warangal, India.

F-EIR Conference 2021 on
Environment Concerns and its Remediation (ECR21)
Chandigarh, India, October 18–22, 2021

Impact of COVID-19 Pandemic Lockdown on the Managerial Aspects of Brick Manufacturing Unit— A Case Study on Post and Present Scenario

P. Murthi[1] and K. Poongodi[2]

ABSTRACT

Every commercial activity has exaggerated due to the ruthless COVID-19 pandemic. Construction sector is one among the hardest affected industry and facing multiple challenges to proceed. The entire civil engineering related activities across the country had halted due to restrictions put in the place by the Central and State Governments of India. Cumulatively, the multiple consequences of the lock down situations would cause the socio-economic condition of the industry. In these circumstances, an insight investigation was carried out by using Likert scale based questionnaire and this study is focused on the managerial condition assessment of the stakeholder of a brick manufacturing chamber, a construction industry driven unit, situated in Warangal, Telangana State, India. The problems with respect to retention of migratory labours, consequence in economic aspects of production process including health aspects of labour and its effect on the present scenario are evaluated and presented.

Keywords: COVID-19, Construction Industry, Brick Manufacturing Kiln, Managerial Issues.

[1] *Presenting & corresponding author.*
Centre for Construction Methods and Materials, SR University, Warangal, India.
E-mail: p.murthi@sru.edu.in
[2] Department of Civil Engineering, SR University, Warangal, India.
E-mail: k.poongodi@sru.edu.in

F-EIR Conference 2021 on
Environment Concerns and its Remediation (ECR21)
Chandigarh, India, October 18–22, 2021

Influence of Coir Fibre and Recycled Aggregate on Bond Strength of Pavement Quality Concrete

K. Poongodi[1], P. Murthi[2] and P. Revathi[3]

ABSTRACT

This investigation program was intended to develop fibre reinforced pavement quality concrete (PQC) with coarse recycled aggregate (CRA) and coir fibre. The work was carried out in two different stages viz. finding the optimum dosage level of CRA and predict the optimum dosage of coir fibre content based on the mechanical properties. M35 grade of PQC was designed for this investigation. The primary objective of this investigation was to evaluate the effect of CRA and coir fibre on bond strength of PQC. The replacement of granite aggregate with CRA by 25% and addition of 1% coir fibre were found as the optimum levels in the concrete composites without compromising the compressive strength and splitting tensile strength. The pull out test was conducted to evaluate the bond strength and found to increase the bond strength at 45% more than the control concrete based on the 0.125 mm slip during the test.

Keywords: *Pavement Quality Concrete, Coir Fibre, Recycled Aggregate, Bond Strength.*

[1] Department of Civil Engineering, SR University, Warangal, India.
E-mail: k.poongodi@sru.edu.in
[2] *Presenting & corresponding author.*
Department of Civil Engineering, SR University, Warangal, India.
Centre for Construction Methods and Materials, SR University, Warangal, India.
E-mail: p.murthi@sru.edu.in
[3] Department of Civil Engineering, Pondicherry Technological University, Puducherry, India.

F-EIR Conference 2021 on
Environment Concerns and its Remediation (ECR21)
Chandigarh, India, October 18–22, 2021

Development of Green Masonry Mortar Using Fine Recycled Aggregate based on the Shear Bond Strength of Brick Masonry

P. Murthi[1], S. Krishnamoorthi[2], K. Poongodi[3] and R. Saravanan[4]

ABSTRACT

The investigation is intended to develop a green masonry mortar using fine recycled aggregate (FRA) for sustainable building construction. The FRA was obtained from the demolished concrete debris and its influence in shear bond strength of brick masonry was evaluated to predict the suitability. The shear bond strength was determined by the masonry triplet specimens. The masonry mortar was prepared by replacing the river sand at the rate of 10%, 20%, 30%, 40% and 50% using FRA. Three different bricks were used for investigation based on its strength. The triplet specimens were casted with 12 mm and 18 mm mortar thickness and the specimens were cured 7, 28 and 56 days. It was evidenced from the investigation that the substitution of recycled aggregate was influenced the shear bond strength of masonry structures and the optimum replacement level was predicted using the relationship between the dosage of recycled aggregate and the shear bond strength.

Keywords: Recycled Fine Aggregate, Masonry Mortar, Shear Bond Strength, Masonry Triplet.

[1] *Presenting & corresponding author.*
Centre for Construction Methods and Materials, SR University, Warangal, India.
E-mail: p.murthi@sru.edu.in
[2] Department of Civil Engineering, Kongu Engineering College, Perundurai, India.
[3] Department of Civil Engineering, SR University, Warangal, India.
[4] Department of Civil Engineering, Navothaya Institute of Technology, Raichur, India.

F-EIR Conference 2021 on
Environment Concerns and its Remediation (ECR21)
Chandigarh, India, October 18–22, 2021

Particleboard Panels Made with Sugarcane Bagasse Waste—An Exploratory Study

Nara Cangussu[1], Patrícia Chaves[2], Welis da Rocha[2] and Lino Maia[3]

ABSTRACT

The reuse of natural fibers, in order to manufacture a new product, is already becoming popular due to the generation of a series of advantages in social areas. Sugarcane bagasse is a set of tangled fibers of cellulose, produced in large quantities due to increased acreage and industrialization of sugarcane resulting from public and private investments in production aimed for the alcohol industry. The aim of this study was to evaluate the feasibility of producing sheet timber manufacture from the sugarcane bagasse, analyzing mechanical strength properties. A form of metal sheet for the molding of 12 specimens based on sugarcane bagasse and industrialized resin was made. Soon after molding, specimens were submitted to a three-point bending test, with the aid of a press. The analysis of the results allowed to conclude that the tensile strength and the modulus of elasticity did not obtain the minimum values recommended by the standard. The tensile strength must be improved to allow panels to be useful for ordinary strength applications.

Keywords: Mechanical Properties, MDP, Recycling, Sugarcane Bagasse, Waste.

[1] CONSTRUCT-LABEST, Faculty of Engineering (FEUP), University of Porto, Rua Dr. Roberto Frias, 4200-465 Porto, Portugal.
FASA, Santo Agostinho Faculty, Av. Osmane Barbosa, 1179-1199 - Jk, Montes Cla-ros-MG, Brazil.
ORCID: 0000-0002-3442-6224. *E-mail:* naracan@gmail.com
[2] FASA, Santo Agostinho Faculty, Av. Osmane Barbosa, 1179-1199 - Jk, Montes Claros-MG, Brazil.
[3] *Presenting & corresponding author.*
CONSTRUCT-LABEST, Faculty of Engineering (FEUP), University of Porto, Rua Dr. Roberto Frias, 4200-465 Porto, Portugal.
Faculty of Exact Sciences and Engineering, University of Madeira, Campus da Penteada, 9020-105 Funchal, Portugal. ORCID: 0000-0002-6371-0179. *E-mail:* linomaia@fe.up.pt

F-EIR Conference 2021 on
Environment Concerns and its Remediation (ECR21)
Chandigarh, India, October 18–22, 2021

Environmental Benefits of Using the Sewage Sludge in the Production of Ceramic Bricks

Nara Cangussu[1], Luana Vasconcelos[2] and Lino Maia[3]

ABSTRACT

The production of sludge from the sewage treatment plants is increasing according to the population increasing and of public policies to improve the sanitation sector. This sludge presents a potential risk to health and the environment, configuring major challenges for sanitation companies regarding treatment and final disposal of this material. The solution is in the circular-economy concept: this sludge presents favorable characteristics to be used as raw-material in the ceramic industry. The present work sought to quantify the environmental impacts related to the atmospheric emissions caused and to the consumption of resources when in the production of bricks 10% of clay is replaced by sewage sludge. Tools of life cycle assessment were used to establish a comparison between the common scenario of the brick production using ceramic mass only from clay and the scenario with the incorporation of 10% of sewage sludge. The results revealed that the incorporation of the sewage sludge has multiple benefits, with the environmental impacts being decreased in all the categories studied.

Keywords: Sewage Sludge, By-product, Circular Economy, LCA, Environmental Impacts, Ceramic Bricks.

---

[1] *Presenting.*
CONSTRUCT-LABEST, Faculty of Engineering (FEUP), University of Porto, RuaDr. Roberto Frias, 4200-465 Porto, Portugal.
Center of Exact and Technological Sciences, Civil Engineering Course, State University of Montes Claros (UNIMONTES), Campus Prof. Darcy Ribeiro, Montes Claros 39401089, Brazil.
ORCID: 0000-0002-3442-6224. *E-mail:* naracan@gmail.com
[2] Center of Exact and Technological Sciences, Civil Engineering Course, State University of Montes Claros (UNIMONTES), Campus Prof. Darcy Ribeiro, Montes Claros 39401089, Brazil.
E-mail: luannavasconcelos@yahoo.com.br
[3] *Corresponding author.*
CONSTRUCT-LABEST, Faculty of Engineering (FEUP), University of Porto, RuaDr. Roberto Frias, 4200-465 Porto, Portugal.
Faculty of Exact Sciences and Engineering, University of Madeira, Campus da Penteada, 9020-105 Funchal, Portugal. ORCID: 0000-0002-6371-0179. *E-mail:* linomaia@fe.up.pt

F-EIR Conference 2021 on
Environment Concerns and its Remediation (ECR21)
Chandigarh, India, October 18–22, 2021

Effect of Vedic Plaster Flooring Over a Rooftop on Room Temperature

Aunj Kumar[1], Shailendra Kumar Jain[2] and Ravinder Kumar Tomar[3]

ABSTRACT

Vedic plaster is newly developed plaster which includes cow dung among other natural materials. Although there are lots of research already done on improvising gypsum and cement plaster, there are very few literatures available on Vedic plaster which is least explored material. Vedic plaster is Gypsum based cow dung plaster. Gypsum ($CaSO_4.2H_2O$) is a mineral composed of calcium sulphate dehydrate. In addition, cow dung contains large amount of protein in it that helps in binding of the bricks and setting of plaster. The setting time of Vedic plaster was observed to be 65 min and set plaster patches did not show any disintegration, popping or pitting. This project is to do a comparative study between normal plaster and Vedic plaster and how Vedic plaster works better than normal plaster in many aspects. It has thermal insulating properties as well as cooling effects during summer and vice versa. It also acts as an air purifier instead of spreading bad odour. It can reduce global issues like global warming and reduction in cost, waste management problems etc. here both the plasters are applied on the top of the roof and will do a comparative study regarding thermal comfort, performance, durability and etc.

Keywords: *Vedic Plaster, Cow Dung, Gypsum, Thermal Comfort.*

[1] *Presenting & corresponding author.*
 Department of Civil Engineering, Amity University, Noida, India.
 E-mail: anujrajput15dec@yahoo.in
[3] Department of Civil Engineering, Amity University, Noida, India.
 E-mail: skjain@amity.edu
[3] Department of Civil Engineering, Amity University, Noida, India.
 E-mail: rktomar@amity.edu

Effect of Reinforcing Graphene Nanoplatelets (GNP) on the Strength of Aluminium (Al) Metal Matrix Nanocomposites

Virat Khanna[1], Vanish Kumar[2] and Suneev Anil Bansal[3]

ABSTRACT

Materials, with high strength applications, are always high in demand for the industry. Metal matrix composites play an important role in providing these materials. In the present study, Aluminium (Al)-graphene nanoplatelets (GNP) nano composites (Al-GNP) were fabricated through the route of powder metallurgy. Pure Al was used as matrix material and GNP (C-750) was used as reinforcement. The process consists of different steps viz dry ball milling followed by compaction and sintering. Compression and hardness test were performed on the formed samples. Four types of composites were fabricated with different proportions of GNP (0.1 wt%, 0.25 wt%, 0.5 wt% and 1 wt%) and the results were compared with pure Al made with the similar process. Morphological characterization like XRD, SEM and FTIR of the formed samples was also performed. It was found that both the hardness and compressive stress were increased up to a certain proportion of GNP in Al-GNP. The maximum hardness achieved was 23 HV with 0.25 wt% GNP which was 98.2% more than the hardness value of pure Al. Similarly, the maximum compressive stress achieved was 109.33 MPa which was 59.18% high than the compressive stress of pure Al.

Keywords: Aluminium, GNP, Powder Metallurgy, Hardness, Compressive Strength.

[1] *Presenting & corresponding author.*
Department of Mechanical Engineering, MAIT, Maharaja Agrasen University, Baddi 174103, India.
E-mail: khanna.virat@gmail.com
[2] National Agri Food Biotechnology Institute (NABI), S.A.S. Nagar, Punjab, 140306, India.
E-mail: vanish.saini01@gmail.com
[3] Department of Mechanical Engineering, MAIT, Maharaja Agrasen University, Baddi 174103, India.
E-mail: suneev@gmail.com

F-EIR Conference 2021 on
Environment Concerns and its Remediation (ECR21)
Chandigarh, India, October 18–22, 2021

Cement After Expiry Date:
Effect in the Concrete Properties

Stéphanie Rocha[1], Cássio Gonçalves[2] and Lino Maia[3]

ABSTRACT

Portland cement has been widely used around the world to produce concrete. When Portland cement is not used correctly, following the normative prescriptions, its use can compromise the aesthetics and the safety of structures. The present research consisted of analyzing the workability of concrete in fresh state and compressive strength in its hardened state when using Portland cement with expired date higher than 90 days defined by the Brazilian standard NBR. For the tests, three cements were selected with different manufacturing dates: November 2018, July 2019 and August 2020. The workability was assessed through the slump test and the compressive strength test at the ages of ages 1, 7, 14 and 28 days. The mix used was for a compressive strength of 25 MPa at the age of 28 days. The slump tests showed divergent values regarding to the expired cements, leading to inconclusive findings, because even with the w/c change, the slump achieved for the 2019 cement-based concrete was lower than that recommended for common concrete. It was concluded that the compressive strength decreased for concretes produced with expired cement.

Keywords: *Cement, Compressive Strength, Concrete, Expired Date, Workability.*

[1] *Presenting & corresponding author.*
CONSTRUCT-LABEST, Faculty of Engineering (FEUP), University of Porto, RuaDr. Roberto Frias, 4200-465 Porto, Portugal.
Civil Engineering Course, Valley of Gorutuba Faculty (FAVAG), Tancredo of Almeida Neves Avenue, 39525-000, Nova Porteirinha, MG, Brazil. ORCID: 0000-0002-0984-4897. *E-mail:* up202010607@g.uporto.pt
[2] Civil Engineering Course, Valley of Gorutuba Faculty (FAVAG), Tancredo of Al-meida Neves Avenue, 39525-000, Nova Porteirinha, MG, Brazil.
[3] CONSTRUCT-LABEST, Faculty of Engineering (FEUP), University of Porto, RuaDr. Roberto Frias, 4200-465 Porto, Portugal.
Faculty of Exact Sciences and Engineering, University of Madeira, Campus da Penteada, 9020-105 Funchal, Portugal. ORCID: 0000-0002-6371-0179. *E-mail:* linomaia@fe.up.pt

F-EIR Conference 2021 on
Environment Concerns and its Remediation (ECR21)
Chandigarh, India, October 18–22, 2021

A Review on Vegetable Oils and Nanofluids as Sustainable Lubricants in Machining

Vishal Yashwant Bhise[1] and Bhagwan Fatru Jogi[2]

ABSTRACT

The lubricants are the vital part in machining which helps to reduce the friction between tool-chip interface and achieve the better machining performance. The conventional lubricants used in the machining are mineral based petroleum oils. It effects on human health during operation and on the environment after the use as it is difficult to dispose. Therefore, there is a need to practice the environmental friendly sustainable cutting fluids. Several studies concluded that the application of sustainable nanofluids imparted better surface finish, tool life, less temperature at tool-chip interface and low cutting forces than the conventional cutting fluids. This paper exhibits a detailed study on the recent practices on machining with the use of various vegetable oils, nanofluids and hybrid nanofluids along with the base fluid as natural oil.

Keywords: *Sustainable Manufacturing, Machining, Vegetable Oils, Nanoparticles, Nanofluids, Hybrid Nanofluids.*

[1] *Presenting & corresponding author.*
PhD Research Scholar, Department of Mechanical Engineering, Dr. Babasaheb Ambedkar Technological University Lonere, Raigad, India. *E-mail:* vishalbhise79@gmail.com

[2] *Corresponding author.*
Professor, Department of Mechanical Engineering, Dr. Babasaheb Ambedkar Technological University Lonere, Raigad, India. *E-mail:* bfjogi@dbatu.ac.in

F-EIR Conference 2021 on
Environment Concerns and its Remediation (ECR21)
Chandigarh, India, October 18–22, 2021

Evaluating Long Term Properties of Concrete Using Waste Beverage Glass

Jagriti Gupta[1], A.S. Jethoo[2] and P.V. Ramana[3]

ABSTRACT

In the world, a large volume of glass (beverage, soda-lime, Toughened, float and mirror glass etc.) is produced for utilizing in the production and manufacturing of versatile articles for harnessing in diverse activities. Out of this huge quantity of glass, a little is recycled and the rest is dumped in open lands. The aim to the study is to produce concrete products that are environment-friendly and it also enlightens the waste glass's enhanced utilization and its application. Therefore, researchers are growing interest in using glass waste as alternative cement and aggregate materials for construction. Therefore, this study characterized the replacement made for cement with the waste beverage glass for various replacements, this characterization was achieved by the produced concrete's durability properties analysis. For the substitution to cement, different ratios have been adopted by weight 0 to 40% at 5% interval. The influence on the mechanical (compression, flexural and abrasion), water absorption and density, permeability and strength loss characterization of waste glass concrete is assessed. The substitution dosages provided to concrete of the glass performs an active role for the concrete characteristics. The experimental investigation reveals some major effects for such beverage glass waste powder replacements that it lowers the density and absorption of water. The reaction of glass waste powder and the hydroxides of calcium i.e. $(Ca(OH)_2)$ is the main reason behind the refinement of the voids, this produces the C-S-H gel termed as hydrates of calcium silicate, and this phenomenon leads to the decrement in the porosity; such reactions are pozzolanic reactions. However, the sulphate resistance of the concrete with an amount of waste glass is higher in terms of compressive strength when the sample is kept in exposer for duration of 90 days.

Keywords: Beverage Glass Waste, Concrete, Durability Properties, ASR Effect, Strength Loss.

[1] *Presenting & corresponding author.*
Department of Civil Engineering, Malaviya National Institute of Technology, Jaipur, India.
E-mail: 2019rce9055@mnit.ac.in
[2] Department of Civil Engineering, Malaviya National Institute of Technology, Jaipur, India.
[3] Department of Civil Engineering, Malaviya National Institute of Technology, Jaipur, India.

F-EIR Conference 2021 on
Environment Concerns and its Remediation (ECR21)
Chandigarh, India, October 18–22, 2021

Effect of Cutting Speed and Feed on Surface Roughness in Dry Turning of Inconel X-750

Vishal Yashwant Bhise[1] **and Bhagwan Fatru Jogi**[2]

Abstract

Dry turning is a sustainable way to carry out the machining but due to low heat precipitation rate of Inconel X 750, the high temperature is generated at the chip-tool interface which further causes built-up-edge formation and consequently creates poor surface finish. As a general practice, sharp edge of cutting tool reduces the vibration and enhances the surface roughness. Accordingly, the use of small nose radius is a better solution to smoothly cut the HRSA materials. Therefore, an attempt has made to study the effect of speed and feed during turning Inconel X-750 under dry environment by using PVD TiAlN coated negative inserts with small nose radius of 0.4 mm. The speed and feed has been applied in three different ranges while depth of cut of 0.5 mm has kept constant. It is observed that at feed rate of 0.1 mm/rev with corresponding speed of 100 m/min, the continuous chips are produced which caused the rubbing action. Hence, excessive localized heat is generated at chip-tool interface and built up edge is formed. Consequently, poor surface finish is produced due to the occurrence of hard turning. Cutting speed and feed rate are significant parameters while studying surface roughness. However, it is observed that the feed rate has great impact on the surface integrity as compared to cutting speed.

Keywords: Inconel X-750, TiAlN Coating, Dry Turning, Surface Integrity.

[1] *Presenting & corresponding author.*
PhD Research Scholar, Department of Mechanical Engineering, Dr. Babasaheb Ambedkar Technological University Lonere, India. *E-mail:* vishalbhise79@gmail.com
[2] *Corresponding author.*
Professor, Department of Mechanical Engineering, Dr. Babasaheb Ambedkar Technological University Lonere, India. *E-mail:* bfjogi@dbatu.ac.in

Lightweight Concrete Containing Fly Ash and Prefabricated Air Bubbles

Suman Kumar Adhikary[1] and Zymantas Rudzionis[2]

ABSTRACT

Every year a large number of waste materials are generated from various sources and waste management becoming a global challenging issue. Besides, the rapid growth of industrialization has resulted in the over-exploitation of natural resources, the reuse of solid waste can be a sustainable approach to overcome those issues. Expanded glass aggregates are produced from waste glass materials and used in the construction sector for their various beneficial aspects. Similarly, fly ash is a by-product of burning pulverized coal used in concrete to improve the properties of concrete. To understand the physical and mechanical properties of lightweight flowable concrete under the various doses of fly ashes ranging from 0% to 30% two series of concrete samples were prepared in this study. Fly ash and prefabricated air bubbles were used as a partial replacement of cement and expanded glass aggregates, respectively. The properties of lightweight flowable concrete including workability and hardened concrete dry density, compressive and flexural strength, microscopic analysis and thermal conductivity were investigated. It has been found from the study that fly ash, and prefabricated air bubbles have a significant impact on lightweight concretes mechanical and physical properties. Almost 37% and 53% reduction in compressive strength was measured for concrete with 30% fly ash; and fly ash with prefabricated air bubbles, respectively.

Keywords: *Lightweight Concrete, Fly Ash, Air Bubbles, Compressive Strength, Thermal Conductivity, Sustainable Concrete.*

[1] Presenting & corresponding author.
  Faculty of Civil Engineering and Architecture, Kaunas University of Technology, Kaunas, LT 44249, Lithuania.
  E-mail: suman.adhikary@ktu.edu
[2] Faculty of Civil Engineering and Architecture, Kaunas University of Technology, Kaunas, LT 44249, Lithuania.
  E-mail: zymantas.rudzionis@ktu.lt

F-EIR Conference 2021 on
Environment Concerns and its Remediation (ECR21)
Chandigarh, India, October 18–22, 2021

Utilization of Waste Granite Powder in Cementitious Composites: Perspectives and Challenges in Poland

Slawomir Czarnecki[1] and Lukasz Sadowski[2]

ABSTRACT

In this research the authors would like to emphasize the need of sustainable waste management due to possible environmental danger in Poland. It is mainly due to maintaining a very high level and even increasing mining of rock raw materials. Along with Poland's accession to the European Union, more and more restrictive regulations on environmental protection were introduced – including those relating to habitat and bird areas, which prevent the use of approx. 500 exploited deposits, which is a reduction of approx. 35% of the possibility of extraction – and those concerning detailed waste management planning. They oblige all producers to design entire production processes so as to improve the environmental preparation of waste for reuse, recycling, recovery and disposal of waste. By wide and sufficient literature survey and statistical data, the analyses of possible waste production is presented. In recent years the emphasis is on obtaining mineral powders that are waste from other industrial processes. On the end of this work the granite powder will be presented as an example of this waste. This waste are more often used as a admixture in cementitious materials, which is also beneficial for eco-friendly way of their utilization. In the article advantages and disadvantages of using granite as by-product in cementitious materials production have been presented. The presentation included the influence of addition of granite at physical, mechanical and chemical properties of cementitious materials.

Keywords: *Waste Mineral Powders, Sustainable Development, Cementitious Materials, Granite.*

[1] *Presenting & corresponding author.*
Dr., Wroclaw University of Science and Technology, Department of Materials Engineering and Construction Processes, Wybrzeze Wyspianskiego 27, 50-370, Wroclaw, Poland. *E-mail:* slawomir.czarnecki@pwr.edu.pl
[2] Prof., Wroclaw University of Science and Technology, Department of Materials Engineering and Construction Processes, Wybrzeze Wyspianskiego 27, 50-370, Wroclaw, Poland. *E-mail:* slawomir.czarnecki@pwr.edu.pl

F-EIR Conference 2021 on
Environment Concerns and its Remediation (ECR21)
Chandigarh, India, October 18–22, 2021

Advances in Pumpability of 3D Printable Concrete Mixes

Spandana Paritala[1], Indira Bathina[2], Mohd Ataullah Khan[3]
and K.N. Chandra Shekaran[4]

ABSTRACT

3D printing of concrete is a rapidly growing technology bringing advantages like freedom of geometry, formwork-free construction, reduced construction time, and wastage. Extrusion-based concrete 3D printing technology is being adopted by the industry owing to its scalability. For such a technology, a printable mix should be capable of being pumped through a distribution network, extruded, and deposited in a layer-by-layer fashion while retaining the shape and integrity of the printed filament. Pumping of the concrete intended for printing is a crucial stage in the digital concrete fabrication process affecting the subsequent stages and the mechanical performance of the hardened elements. Rheological behavior and mix design for the pumping of conventional concrete is well established over the last couple of decades. This study provides a review of significant rheological properties and the effect of various material mixes on pumping of conventional as well as printable concrete, and a general comparison between them. The significant rheological properties affecting the printability of concrete are static yield stress, plastic viscosity, thixotropy, and open time. The material mixes constituents influencing the pumpability of concrete, identified here are the type of binder(s), the inclusion of nano-fillers, aggregate shape and grading, aggregate-to-cement ratio. In addition, the geometric and material configuration of the printer also influences pumpability. The mechanisms of pumping conventional and printable concrete through pipes is discussed. Different test methods adopted to measure the rheological properties and/or pumpability of conventional and 3D printable concrete are presented. Optimization of the rheological framework of printable concrete required for appreciable buildability and hardened properties along with successful pumpability is highlighted. The range of yield stress, plastic viscosity, and thixotropy is tabulated for 3D printable mixes.

Keywords: Concrete 3D Printing, Pumpability, Supplementary Cementitious Materials, Extrusion-Based Printing, Rheology, Thixotropy.

[1] *Presenting author.* Undergraduate student, Department of Civil Engineering, Rajiv Gandhi University of Knowledge Technologies, Nuzvid, India. *E-mail:* spandanaparitala@gmail.com

[2] Undergraduate student, Department of Civil Engineering, Rajiv Gandhi University of Knowledge Technologies, Nuzvid, India. *E-mail:* indirabathina@gmail.com

[3] *Corresponding author.* Assistant Professor, Department of Civil Engineering, Mahindra University, Ecole Centrale College of Engineering, Hyderabad, India. *E-mail:* mohdataullah.khan@mahindrauniversity.edu.in

[4] Chief Consultant, Indraprasta Consultants, Hyderabad, India. *E-mail:* Kncipc@gmail.com

F-EIR Conference 2021 on
Environment Concerns and its Remediation (ECR21)
Chandigarh, India, October 18–22, 2021

Adsorptive Removal of Diclofenac Sodium Using Graphene Oxide: An Insight to the Kinetic and Thermodynamic Aspects

Sumit K. Singh[1], Manish Kumar[2] and Hema Setia[3]

ABSTRACT

Increased consumption of pharmaceuticals & personal care products (PPCPs) has led to introduction of highly stable pollutants. Pharmaceuticals particularly may be taxing to remove using existing technologies and poses a great threat to humans and the entire ecosystems and necessitates the development of robust technologies for the removal of PCPs. Technological advancements in this domain over recent decades have led to the evolution of highly efficient materials adept for degrading a wide range of emerging pollutants using photocatalysis and advanced oxidation processes but leads to the formation of stable byproducts and unnecessary oxidative stress within the water matrices. The current study attempts to investigate adsorption using graphene oxide as noble method for removal of diclofenac sodium as a model pharmaceutical contaminant while focusing synthesis and purification and characterization of graphene oxide. Different characterization techniques namely XRD, FTIR and Electron microscopy have been used the determine composition, geometry, and morphology of the synthesized graphene oxide. Adsorptive behavior is studied in terms of adsorbent loading, adsorbate concentration and operational pH. The adsorption isotherms and kinetics for the GO-DCF-Na system is analyzed after fitting the data obtained to suitable models. A thermodynamic analysis is also performed to evaluate different thermodynamic aspects and the feasibility of the adsorptive process.

Keywords: Adsorption Isotherm, Diclofenac Sodium, Emerging Contaminants, Graphene Oxide.

[1] *Presenting author.*
Biotechnology, University Institute of Engineering and Technology, Panjab University, Chandigarh, India.
E-mail: sumit.pallava@hotmail.com
[2] Biotechnology, University Institute of Engineering and Technology, Panjab University, Chandigarh, India.
E-mail: mk12111999@gmail.com
[3] *Corresponding author.*
Biotechnology, University Institute of Engineering and Technology, Panjab University, Chandigarh, India.
E-mail: hemasetia@pu.ac.in

F-EIR Conference 2021 on
Environment Concerns and its Remediation (ECR21)
Chandigarh, India, October 18–22, 2021

Small Percentage Reinforcement of Carbon Nanotubes (CNTs) in Epoxy (Bisphenol-A) for Enhanced Mechanical Performance

Suneev Anil Bansal[1], Virat Khanna[1], Twinkle[2],
Amrinder Pal Singh[3] and Suresh Kumar[4]

ABSTRACT

Epoxies, in its various forms, find wider applications in the field of automobile, aircrafts, marine and other structures. The recent development of carbon based nano particles like carbon nanotubes (CNTs), as reinforcement, has increased the scope to enhance mechanical properties of epoxy based composites. The present report is an attempt to understand the effects of very small percentage reinforcement of CNTs (i.e. 0.25 wt.%) in epoxy on mechanical properties of nanocomposite. The solution mixing method was used to synthesize the nanocomposite. The CNTs were dispersed in a solvent using the probe sonicator. After complete dispersion of the CNTs, the epoxy was added under constant stirring condition. To cure the epoxy, amine based curing agent was used at room temperature for 24 hours. Post curing was done for 2 hours at 80°C. The samples were prepared in aluminum moulds. Morphology of nanocomposite was studied with scanning electron microscopy (SEM). Mechanical properties of nanocomposites were characterized using nano indentation technique. The results of elastic modulus and hardness showed enhancement in mechanical properties of the synthesized nanocomposites.

Keywords: CNTs, Epoxy, Mechanical Properties, Nanocomposite, Nano Indentation.

[1] *Presenting & corresponding author.*
Department of Mechanical Engineering, MAIT, Maharaja Agrasen University, Baddi 174103, India.
E-mail: suneev@gmail.com
[2] *Presenting author.*
Department of Applied Sciences, UIET, Panjab University, Chandigarh 160014, India.
[3] Department of Mechanical Engineering, UIET, Panjab University, Chandigarh 160014, India.
[4] Department of Applied Sciences, UIET, Panjab University, Chandigarh 160014, India.

F-EIR Conference 2021 on
Environment Concerns and its Remediation (ECR21)
Chandigarh, India, October 18–22, 2021

Statistical Analysis of Hooked-End Steel Fibers in Concrete

Mandala Venugopal[1], Jayaprakash Vemuri[2] and Mohd Ataullah Khan[3]

ABSTRACT

Concrete, owing to its brittle nature is weak in tension and therefore, a significant amount of published research has focused on the addition of fibers to improve the flexural performance of concrete. The addition of fibers in a concrete matrix improves its overall performance in terms of strength, ductility, and durability by the mechanism of restraining crack width. Steel fibers are found to be the most efficient in improving the flexural strength of concrete. The design mix for Steel Fiber-Reinforced Concrete (SFRC) is chosen mostly based on the trial mixes. To aid the mix design of SFRC, this study attempts to provide predictive models using machine learning techniques such as non-linear regression on a database of 146 samples, collated from literature. 80% of the randomly chosen samples are used for training the model and the remaining 20% was used to test it. The Root Mean Squared Error (RMSE), Mean Absolute Error (MAE), the statistical significance of the model coefficients, and the coefficient of determination values are adopted as criteria to check the performance of the model on both the training and test data. Interaction terms such as fiber factor and number of fibers within a volume were inferred from literature to improve the efficacy of the predictive modeling. Two non-linear models for the prediction of compressive and flexural strength of hooked-end SFRC are proposed and validated. Using the proposed models, the variations in compressive and flexural strength with the key parameters were plotted and discussed. With a decrease in the water-to-binder ratio, there was a significant improvement in compressive strength from 55 MPa to 85 MPa at an aspect ratio of 60. A significant improvement of about 200% to 300% in flexural strength was noted with an increase in fiber factor from 0 to 150.

Keywords: *Hooked-end Fibers, SFRC, Regression Analysis, Volume of Fibers, Aspect Ratio, Fiber Factor.*

---

[1] *Corresponding author.*
PhD Student, Department of Civil Engineering, Mahindra University, Ecole Centrale College of Engineering, Hyderabad, India. *E-mail:* venugopal20pcie007@mahindrauniversity.edu.in
[2] Assistant Professor, Department of Civil Engineering, Mahindra University, Ecole Centrale College of Engineering, Hyderabad, India. *E-mail:* jayaprakash.vemuri@mahindrauniversity.edu.in
[3] Assistant Professor, Department of Civil Engineering, Mahindra University, Ecole Centrale College of Engineering, Hyderabad, India. *E-mail:* mohdataullah.khan@mahindrauniversity.edu.in

F-EIR Conference 2021 on
Environment Concerns and its Remediation (ECR21)
Chandigarh, India, October 18–22, 2021

A Predictive Model for Fly Ash and Slag Blended Geopolymer Concrete

S. Kailash Kumar[1], Mohd Ataullah Khan[2] and T. Visalakshi[3]

ABSTRACT

A predictive model for the compressive strength of fly ash and slag blended GPC is developed using non-linear regression analysis as a machine learning tool, on a database aggregated from the literature. Statistical investigations on the influence of parameters, namely sodium silicate-to-hydroxide ratio, activator-to-binder ratio, Si-to-Al ratio, aggregate-to-binder ratio, Ca-to-Si ratio, Super plasticizer dosage, the molarity of NaOH, curing temperature, water-to-geopolymer-solids ratio on the compressive strength are carried out while developing the model. 75% of the data are randomly chosen to train the model while the remaining 25% was used to test the efficacy of the proposed model. Root Mean Squared Error of the predictions on the training and testing datasets, and R-squared value of the model based on training data are used as criteria to select the model. The statistical investigation on the gathered dataset shows that the compressive strength is significantly affected by the combined effect of the proportion of sodium silicate-to-hydroxide ratio, activator-to-binder ratio, Si-to-Al ratio, Ca-to-Si ratio, and the curing temperature. Contour plots obtained from the developed model are used to predict the relationship between the input parameters and compressive strength of Fly ash and Slag blended GPC. From the contour plots, it was found that with the increase in Ca-to-Si ratio and curing temperature the compressive strength was increasing while it was decreasing with the increase in Si-to-Al ratio, sodium silicate-to-hydroxide ratio, and activator-to-binder ratio. The results of proposed models predict that the Fly ash and Slag blended GPC can be fabricated without the need of high thermal curing with the compressive strength in the range of (70 MPa to 90 MPa), provided that the following range is adopted: the ratio of sodium silicate to sodium hydroxide (1.5–2.5), activator-to-binder ratio (0.2–0.4), Si-to-Al ratio (1.4–1.8), Ca-to-Si ratio (0.4–0.6), and curing temperature (30°C to 60°C).

Keywords: Nonlinear Regression, Sustainable, Slag, Compressive Strength, Fly ash, Machine Learning.

[1] Research Scholar, Department of Civil Engineering, Mahindra University, Ecole Centrale School of Engineering, Hyderabad, India. *E-mail:* kailash20pcie010@mahindrauniversity.edu.in
[2] *Presenting & corresponding author.*
Professor, Department of Civil Engineering, Mahindra University, Ecole Centrale School of Engineering, Hyderabad, India. *E-mail:* mohdataullah.khan@mahindrauniversity.edu.in
[3] Assistant Professor, Department of Civil Engineering, Mahindra University, Ecole Centrale School of Engineering, Hyderabad, India. *E-mail:* visalakshi.talakokula@mahindrauniversity.edu.in

F-EIR Conference 2021 on
Environment Concerns and its Remediation (ECR21)
Chandigarh, India, October 18–22, 2021

Physical and Mechanical Properties of Concrete Made with Glass Sand

Filipe Figueiredo[1], Rayssa Renovato Reis[2] and Lino Maia[3]

ABSTRACT

It is important promoting less generation of urban waste not only for the economy of raw materials, but also for the lesser deposition of waste in land-fills. Glass is among the most frequently used materials in packaging and is the easiest to be recycled. Being aware that the construction industry is one of the largest consumers of natural resources in the world, the present study deals with the possibility of replacing part of the fine aggregate of concrete by glass sand. Concrete was produced with the conventional aggregates and then the fine aggregate was partially replaced by glass sand for the percentages of 10, 15 and 20%. The experimental study sought to evaluate the physical and mechanical properties. The slump, compressive strength, tensile strength and water absorption by capillarity and void index testing were measured. Results showed that when replacing the fine aggregate by 10, 15 and 20% of glass sand the results are satisfactory, especially for the substitution of 20% of glass sand which presented an increase of 18% in the compressive strength.

Keywords: Glass Sand, Fine Aggregate, Concrete.

[1] CONSTRUCT-LABEST, Faculty of Engineering (FEUP), University of Porto, RuaDr. Roberto Frias, 4200-465 Porto, Portugal.
UFGD, Federal University of Grande Dourados, Rua João Rosa Góes, nº 1761, Vila Progresso, Dourados,79825-070, MS, Brazil. ORCID: 0000-0003-3191-0683. *E-mail:* filipefigueiredo@ufgd.edu.br

[2] UFGD, Federal University of Grande Dourados, Rua João Rosa Góes, nº 1761, Vila Progresso, Dourados,79825-070, MS, Brazil.

[3] *Presenting & corresponding author.*
CONSTRUCT-LABEST, Faculty of Engineering (FEUP), University of Porto, RuaDr. Roberto Frias, 4200-465 Porto, Portugal.
Faculty of Exact Sciences and Engineering, University of Madeira, Campus da Pen-teada, 9020-105 Funchal, Portugal. ORCID: 0000-0002-6371-0179. *E-mail:* linomaia@fe.up.pt

F-EIR Conference 2021 on
Environment Concerns and its Remediation (ECR21)
Chandigarh, India, October 18–22, 2021

High Levels of Nitrate in Well Waters of Saipem Ward, Candolim, Goa

Roshan Ratnakar Naik[1] and Mrunal Rushikesh Phadke[2]

ABSTRACT

Saipem ward in Candolim is located below a plateau onto which the Pilerne industrial estate is situated and locals have been complaining of industrial waste and sewage polluting the wells since last several years. We analyzed water samples from five different households to determine water quality as part of our community outreach programme. We analyzed dissolved oxygen (DO), biological oxygen demand (BOD), chemical oxygen demand (COD), nitrate, total dissolved solids (TDS), total suspended solids (TSS) and most probable number (MPN) test to detect coliform in well water samples. The BOD, COD, TDS, TSS were largely within permissible limits. The nitrate content was greater than 45 mg/l (acceptable limit), thus indicating water unfit for consumption. The coliform index was also very high and variable from 33 to greater than 2400 coliforms/100 ml. We carried out subsequent confirmatory tests using differential media such as eosin methylene blue agar, endo agar and observed gram negative cocci or rods with blue-black colonies but no presence of *E. coli* (common fecal coliform) suggesting water contamination by industrial or agricultural discharge seeping into wells. We also carried out nitrate reduction tests with isolated colonies and observed presence of nitrate reducing enzyme which converts the nitrate (NO_3^-) to nitrite (NO_2^-) ions in water. We will discuss bioaugmentation strategies to further convert NO_2^- to N2 gas for bioremediation purposes.

Keywords: Pollution, Nitrate, Most Probable Number, Coliform, Industrial Waste, Bioremediation.

---

[1] *Presenting & corresponding author.*
Department of Biotechnology, Dhempe College of Arts and Science, Panaji, India.
E-mail: roshan@dhempecollege.edu.in
[2] Department of Biotechnology, Dhempe College of Arts and Science, Panaji, India.
E-mail: mrunal@dhempecollege.edu.in

F-EIR Conference 2021 on
Environment Concerns and its Remediation (ECR21)
Chandigarh, India, October 18–22, 2021

A Short Review on Environmental Impacts and Application of Iron Ore Tailings in Development of Sustainable Eco-Friendly Bricks

H.K. Thejas[1] and Nabil Hossiney[2]

ABSTRACT

Increased mining activity of iron ore has led to the generation of voluminous wastes of various nature, especially during the different stages of its extraction and production. The improper disposal of such waste causes negative impact on the environment. One such waste which is generated during the beneficiation process of iron ore is waste iron ore tailings, which is also termed as IOT. Further, dumping of IOT on open ground creates huge dumping sites. This dumping sites have been a concern to the environment and human population in its close vicinity. Therefore, a need to effectively use IOT has become one of the subjects of interest for many researchers. This article provides a short review of environmental problems caused due to improper disposal of IOT, and also reviews on the reuse methods of IOT in the construction sector, which helps to alleviate the environmental pollution associated with improper disposal of IOT. Furthermore, reuse of IOT in construction sector reduces the exploitation of the virgin materials for production of construction material, and thus reducing depletion of natural resources. Based on the existing literatures and findings it was observed that the use of IOT to develop stable building blocks using unconventional methods showed great potential and improved performance, when compared with conventional materials such as clay fired bricks.

Keywords: Iron Ore Tailings, Environment, Sustainable, Eco-friendly, Bricks.

---

[1] *Presenting & corresponding author.*
Assistant Professor, Department of Civil Engineering, CHRIST (Deemed to be University), Bangalore, India.
E-mail: thejas.hk@christuniversity.in
[2] Associate Professor, Department of Civil Engineering, CHRIST (Deemed to be University), Bangalore, India.
E-mail: nabil.jalall@christuniversity.in

State-of-the-Art Analysis on Causes and Repair of Cracks in Rigid Pavement

Shaheen Sulaiman[1], Pavan Chandrasekar[2], V. Poornima[3], R. Venkatasubramani[4] and V. Sreevidya[5]

ABSTRACT

Failures in the rigid pavement can be observed due to the formation of cracks on the pavement surface. Due to continuous moving loading, quality of materials, climatic conditions, and method of construction, pavement undergoes several types of distress and in general, cracks appear on the surface. These cracks may lead to the formation of adjacent cracks, which have to be repaired. This paper reviews the studies on the formation of cracks in detail and focuses on the method used to analyze these types of cracks. The basic distress in the rigid pavement is fatigue cracking, shrinkage cracking, pumping, faulting, durability cracking, blow-ups, alkali-aggregate reaction, linear cracking, spalling, polished aggregates, punch out, corner breaks, pop-outs, and deterioration of joint load transfer system. Repair techniques used in the rigid pavement are crack filling, diamond grinding, stitching, dowel bar retrofitting, partial-depth repair, full-depth repairs, crack sealing and joint repair. Some remedial measures like the addition of fibres and bacterial solution while placing rigid pavement may lead to the reduction in cracks. The paper also presents a detailed review of various methods available for the repair of cracks and discusses the future scope of research on the same.

Keywords: Rigid Pavement, Cracks, Repairing Methods.

[1] Student, Department of Civil Engineering, Amrita School of Engineering, Coimbatore, Amrita Vishwa Vidyapeetham, India. *E-mail:* shaheensulaiman3@gmail.com

[2] *Presenting & corresponding author.*
Student, Department of Civil Engineering, Amrita School of Engineering, Coimbatore, Amrita Vishwa Vidyapeetham, India. *E-mail:* pavanchandrasekar7@gmail.com

[3] *Corresponding author.*
Assistant Professor, Department of Civil Engineering, Amrita School of Engineering, Coimbatore, Amrita Vishwa Vidyapeetham, India. *E-mail:* v_poornima@cb.amrita.edu

[4] Professor, Department of Civil Engineering, Dr Mahalingam College of Engineering and Technology, Coimbatore, India. *E-mail:* rvs_vlb@yahoo.com

[5] Professor, Department of Civil Engineering, Sri Krishna College of Technology, Coimbatore, India. *E-mail:* v.sreevidya@skct.edu.in

Study on Potency of Anti-Corrosion Degree of Waterproofing Agent and Plant Extract to Steel Rebar in Concrete

Madan Somai[1], Ajaya Giri[1], Akash Roka[1] and Jagadeesh Bhattarai[2*]

ABSTRACT

The reinforcement of steel rebar in concrete infrastructures has been practiced for long times mostly to improve tensile strength and to increase their service time. However, shortcomings associated with reinforced steel corrosion damages in concrete infrastructures becoming the matter of corrosion researchers in recent years, mostly from safety and economic aspects. In general, the steel-reinforced concrete infrastructures are porous, and thence water as well different atmospheric pollutant gases penetrate it causing the corrosion damages of the steel rebars, which shorten the prolonged existence of the concrete structures. To avoid such corrosion damage problems of the steel-reinforced concrete infrastructures, cement-based waterproofing admixture is practiced with aiming to decrease the capillary porosity of the concrete infrastructures, making it less porous and more resistant to aggressive agents. Besides, in recent years, there are using green-based corrosion inhibitors acquired from different parts of plant extracts for combating the damage problems of steel rebars in concrete matrixes.

Taking into consideration these facts, the present study was aimed to compare the efficacy and prosperity of two commercially used waterproofing chemicals and methanol extract of *Mangifera indica* leaves for controlling the corrosion susceptibility of concrete rebar using a half-cell corrosion potential measurement method following the ASTM C876-91 standard. Based on the experimental results, it reported that the green-based extract of *M. indica* leaves and waterproofs have found to be high anti-corrosion characteristics and could be recommended for the protection of the reinforced rebars in concrete infrastructures. Although, the corrosion controlling performance degree of 1000 ppm and 2000 ppm *M. indica* extracts on the steel rebars in concrete was higher than the additions of alike concentration of both the waterproofs based on the shifting of corrosion potential values to a more positive than −126 mV (SCE).

Keywords: Additive, Degradation, Electrochemistry, FTIR, Reinforcement.

---

[1] Central Department of Chemistry, Tribhuvan University, Kirtipur, Nepal.
[2] *Presenting & corresponding author.*
Central Department of Chemistry, Tribhuvan University, Kirtipur, Nepal. *E-mail:* *bhattarai_05@yahoo.com

F-EIR Conference 2021 on
Environment Concerns and its Remediation (ECR21)
Chandigarh, India, October 18–22, 2021

Fine Aggregates of CDW: Feasibility of Its Application in the Manufacture of Mortars for Laying

Nara Cangussu[1], Emanuel Silva[2], Rogério Borges[2] and Lino Maia[3]

ABSTRACT

One of the problems faced by the sector of civil construction currently is related to the difficulty to obtain environmental licenses for the extraction of natural aggregates, in due to the environmental impacts caused by such activity. The objective of this study is to propose the replacement of natural sand and gravel with aggregates of construction and demolition waste (CDW) in the production of mortar for laying. For the characterization of the mortars were determinate the consistency index, water retention, specific gravity, tensile strength in flexion and compressive strength. The CDW fine aggregates replaced the natural sand, in mixtures of mortar without to prejudice mortar properties.

Keywords: Recycled Aggregates, CDW, Concrete, Mortar.

[1] *Presenting.*

CONSTRUCT-LABEST, Faculty of Engineering (FEUP), University of Porto, RuaDr. Roberto Frias, 4200-465 Porto, Portugal.

FASA, Santo Agostinho Faculty, Av. Osmane Barbosa, 1179-1199 - Jk, Montes Claros-MG, Brazil.

ORCID: 0000-0002-3442-6224. *E-mail:* naracan@gmail.com

[2] FASA, Santo Agostinho Faculty, Av. Osmane Barbosa, 1179-1199 - Jk, Montes Claros-MG, Brazil.

[3] *Corresponding author.*

CONSTRUCT-LABEST, Faculty of Engineering (FEUP), University of Porto, RuaDr. Roberto Frias, 4200-465 Porto, Portugal.

Faculty of Exact Sciences and Engineering, University of Madeira, Campus da Penteada, 9020-105 Funchal, Portugal. ORCID: 0000-0002-6371-0179. *E-mail:* linomaia@fe.up.pt

F-EIR Conference 2021 on
Environment Concerns and its Remediation (ECR21)
Chandigarh, India, October 18–22, 2021

Tech-En-Econ-Energy-Exergy-Matrix (T4EM) Observations of Evacuated Solar Tube Collector Augmented Solar Desaltification Unit: A Modified Design Loom

Ashok Kumar Singh[1] and Samsher[2]

ABSTRACT

In the present study, a sophisticated novel design of solar desaltification unit (SDU) with evacuated solar tube collector integrated unique combination of modified compound parabolic concentrators (MCPC) has been analyzed Tech-En-Econ-Energy-Exergy-Matrix (T4EM) observations under the specific meteorological conditions of New Delhi, India. The proposed model undergoes a novel design having a boosting performance due to its unique design considerations (i.e. utilization of partially diffused solar radiations through segmented parabolic collector towards the identically exposed EATC) and associated arrangement of oriented compound parabolic concentrators (OCPC) and cusp compound parabolic concentrators (CCPC) (accumulatively the MCPC) integrated to the South faced solar desaltification unit. The current approach emphasizes the EATC utility and performance by utilizing 200% of solar radiation uniformly through its periphery and enhancing the thermo siphons working principle appreciably. Further, the utilization of optimum water mass and number of EATCs enables the proposed system comparatively better in performance than of the single slope SDU and double slope-SDU or with the nearby combinations in terms of yield, economy (environ, energy, exergy), energy matrices, design, system efficiency, life cycle conversion efficiency, carbon mitigate, emission, and credits, yield cost per liter, etc. The explored results are well compared and validated with the existing researches and found appreciable increment in the performance of the proposed system in terms of the aforementioned categories. The entire work has been sensibly concluded with the positive recommendations ahead.

Keywords: MCPC, Solar Desaltification, EATC, Efficiency, Energy Matrices, Economic.

---

[1] *Presenting & corresponding author.*
Department of Mechanical Engineering, Delhi Technological University, Delhi, India.
Mechanical Engineering Department, Galgotias College of Engineering and Technology, Greater Noida, India.
E-mail: agashok26@gmail.com
[2] Department of Mechanical Engineering, Delhi Technological University, Delhi, India.

Removal of Nickel from Aqueous Solution by Using Corncob as Adsorbent

Vishal Chaudhari[1] and Manish Patkar[2]

ABSTRACT

Heavy metal deposition in the environment is increasing day by day as a result of the rapid growth of the industrial sector and massive anthropogenic activities. Among numerous heavy metals, nickel is the 5th greatest copious component on the planet. Nickel is a non-biodegradable, brittle, ductile, and toxic heavy metal ion that is found in drinking water and wastewater. Though numerous conventional treatments are available to remediate nickel-contaminated water, including electrochemistry, chemical precipitation, flocculation, ion exchange, coagulation, membrane filtration, and adsorption, each has some or all of the following drawbacks in terms of cost and efficiency in removing metal ions from water. Among all methods, the adsorption method was found to be the most cost-effective and efficient at removing metal ions from water. Adsorption is a highly efficient and cost-effective method of extracting nickel ion from water that involves surface phenomena. The primary goal of this paper is to identify the best adsorbent for a batch study in extracting nickel ion from aqueous solution. Corncob is used as a low-cost agro-based adsorbent in this study to remove nickel ion from an aqueous medium. To test the suitability of locally available low-cost corncob for nickel-contaminated water, a batch adsorption method was used. The effect of several major adsorption operating parameters have been considered and optimized. It was discovered that the optimum adsorbent dose was 1.5 g/50 ml, that equilibrium was achieved in 120 minutes, and that better adsorption at pH 8 was obtained. Freundlich, Langmuir, and Temkin isotherm models were used to correlate sorption data. The result demonstrates that the Langmuir adsorption model fit best when compared to other adsorption isotherm models, with R^2 = 0.9781 for the given adsorbent. Adsorption kinetics are discussed in order to investigate the rate of nickel ion uptake on the adsorbent. A kinetic study revealed that pseudo-second-order was better fitted best.

Keywords: Adsorption, Batch Study, Corncob, Nickel.

[1] *Presenting & corresponding author.*
Research Scholar, Civil Engineering Department, Oriental University, Indore, India.
E-mail: chaudhariv85@gmail.com
[2] Associate Professor, Civil Engineering Department, Oriental University, Indore, India.
E-mail: manish.patkar@gmail.com

F-EIR Conference 2021 on
Environment Concerns and its Remediation (ECR21)
Chandigarh, India, October 18–22, 2021

Study of Antimony Doping in Silicene Nanoribbons in the Presence of an External Electric Field

Hoang Van Ngoc[1]

ABSTRACT

With the continuous development of materials science, science and nanotechnology contribute significantly to the creation of valuable products. Nanomaterial creates products that are smaller, lighter, cheaper, more efficient, and more fuel-efficient. Silicene is a two-dimensional structure made of silicon atoms with a honeycomb hexagonal configuration. Silicene is not a flat structure like graphene, it has buckling. Silicene nanoribbons (SiNRs) are low-dimensional structures built from silicene with chemically modified edges. The electronic properties of SiNRs also depend on the modification elements at the two edges, the SiNRs studied here have hydrogen-modified edges. The antimony element will be doped in the SiNRs, and the system is placed in an external electric field of 0.3 V/ Angstrom. There are three doping configurations studied here namely top-configuration, valley-configuration, and 100%-configuration. The configurations are stable after optimization by density function theory (DFT), and the formation energies of the systems will also be calculated. The energy band structure, density of states of the configurations will be plotted, analyzed, and compared. The structures after doping are all semi-metallic, the band gap has disappeared. The contributions of partial states are also studied, which depends on the density of particles per unit cell. The appearance of an external electric field changes the energy band structure and state density of the system.

Keywords: *Silicene Nanoribbons, Antimony Doping, External Electric Field, Top-Configuration, Valley-Configuration, 100%-Configuration.*

[1] *Presenting & corresponding author.*
Institute of Applied Technology, Thu Dau Mot University, Thu Dau Mot city, Binh Duong province, Vietnam.
E-mail: ngochv@tdmu.edu.vn

F-EIR Conference 2021 on
Environment Concerns and its Remediation (ECR21)
Chandigarh, India, October 18–22, 2021

Sewage Sludge Ash for Soil Stabilization: A Review

Aparna R. Pillai[1]

ABSTRACT

Rapid industrialization leads to an increase in the quantity of combustion deposits like ashes. Currently, most of these ashes are disposed of in landfills. However, in India, the facilities for disposing these waste is very limited. Many states in India doesn't have a proper disposal mechanism for such wastes, and thus the improper dumping of waste will affect the natural ecosystem. The waste management by reuse or recycling is crucial in the present-day scenario. With the fast growth in the volume of sewage sludge obtained from sewage treatment plants, it is crucial to reuse and reduce sludge. Sewage sludge ash (SSA) obtained from the burning of sewage sludge has significant dissimilarities with the fly ash/bottom ash in terms of chemical, physical and cementitious characteristics. Soil stabilization is indispensable due to the increase in the construction of structures. The studies dealing with soil stabilization has more focus on the use of chemical additives which are expensive and unsafe. Improving the ground using waste materials is one of the most essential and cost-effective methods. The commonly used additives such as cement and lime for stabilizing poor ground can be replaced with waste material, leading to a sustainable and environment friendly approach for discarding the industrial waste. This paper briefly describes the production, properties and application of SSA in civil engineering even though the focus was on the utilization of SSA for soil stabilization. Studies have already reported that SSA possesses pozzolanic properties and has the potential to stabilize the soft ground. This paper also addresses a broad review of the index and engineering properties of soil improved with SSA.

Keywords: Sewage Sludge Ash, Waste Management, Soil Improvement, Index Properties, Engineering Properties.

[1] *Presenting & corresponding author.*
Adhoc Faculty, Civil Engineering Department, National Institute of Technology Calicut, India.
E-mail: draparnaiitm@gmail.com

F-EIR Conference 2021 on
Environment Concerns and its Remediation (ECR21)
Chandigarh, India, October 18–22, 2021

Progressive Collapse Resistance of Reinforced Concrete Frame Structures Subjected to Column Removal Scenario

Pawan Kumar[1], Samrat Lavendra[2] and T. Raghavendra[3]

ABSTRACT

The progressive collapse of a building occurs due to an unusual loading condition or when there is a sudden loss of the primary structural element. This could lead to a partial collapse or entire collapse of the building due to the failure of one or more structural elements. This study investigates the progressive collapse demand of the RC building by linear static analysis and vertical displacement responses using non-linear dynamic analysis. ETABS software is used for corner column removal, edge column removal, penultimate column removal and internal column removal scenarios. The RC buildings considered in this investigation are G+6, G+8 and G+10 storey. The results showed that progressive collapse demand value of the bottom end of adjacent columns were found to be safe for all damage cases. However, the top end of the adjacent columns was found to be unsafe. The most critical damage case was found to be the penultimate column removal case that possessed highest PCD values. Finally, the vertical displacement of time history results is represented graphically and later the results of time history analysis is compared with linear static analysis results.

Keywords: *Column Removal Scenario, Progressive Collapse, Progressive Collapse Demand, Reinforced Concrete Frame, Vertical Displacement.*

[1] *Presenting & corresponding author.*
Post Graduate, Structural Engineering, Department of Civil Engineering, RV College of Engineering, Visvesvaraya Technological University, India. *E-mail:* pawank1617@gmail.com

[2] Post Graduate, Structural Engineering, Department of Civil Engineering, RV College of Engineering, Visvesvaraya Technological University, India. *E-mail:* samratlavendra01@gmail.com

[3] Associate Professor, Department of Civil Engineering, RV College of Engineering, Visvesvaraya Technological University, India. *E-mail:* raghavendrat@rvce.edu.in

Materials and Resources in Green Building Rating Systems: Filling the Environmental Gap in Energy Performance Certificates

Alexandre Soares dos Reis[1], **Marta Ferreira Dias**[2]
and Alice Tavares Costa[3]

ABSTRACT

Energy performance certificates (EPCs) are an integral part of the Energy Performance of Buildings Directive (EPBD) and should have an essential role in enhancing the energy performance of buildings. For this reason, further improvements of EPCs might be necessary. Nowadays, in Portugal, contractors may refurbish buildings without proving any impact on the environment of the components used. This emphasizes the need to track the environmental issues previously and during the rehabilitation of buildings' decisions. As stated in the EPBD, Portugal developed a long-term renovation strategy to decarbonize the building stock by 2050, the so-called ELPRE. According to ELPRE, minimum performance requirements can be complemented by Green Building Rating Systems (GBRS) like Leadership in Energy and Environmental Design (LEED), Building Research Establishment Environmental Assessment Methodology (BREEAM), and *Liderar pelo Ambiente* (LiderA). We found that the GBRS can establish the requirements to increase the demand for building components that minimize the environmental impacts of its use. Current EPCs have, in some cases, limited reliability, compliance, market penetration, or acceptance by the users. A new approach must provide a sustainable and more reliable service to end-users, informing them also about the environmental impacts and not only about the energy-related indexes. Based on LiderA, we propose a new label

[1] *Presenting & corresponding author.*
BSc, Research Unit on Governance, Competitiveness and Public Policies (GOVCOPP), Department of Economics, Management, Industrial Engineering and Tourism (DEGEIT), University of Aveiro, Aveiro, Portugal. *E-mail:* alexandre.soares.reis@ua.pt

[2] PhD, Research Unit on Governance, Competitiveness and Public Policies (GOVCOPP), Department of Economics, Management, Industrial Engineering and Tourism (DEGEIT), University of Aveiro, Aveiro, Portugal. *E-mail:* mfdias@ua.pt

[3] PhD, Research Center for Risks and Sustainability in Construction (RISCO), Department of Civil Engineering (DECivil), Centre for Research in Ceramics and Composite Materials (CICECO), University of Aveiro, Aveiro, Portugal. *E-mail:* tavares.c.alice@ua.pt

based on environmental performance, looking beyond energy. Next-generation EPCs could consider the proposed environmental label, filling the identified environmental gap.

Keywords: *Energy Performance Certificates, Environmental Impacts, Green Building Rating Systems, Local Production, Sustainably Sourced Wood, Responsibly Sourced Building Materials.*

F-EIR Conference 2021 on
Environment Concerns and its Remediation (ECR21)
Chandigarh, India, October 18–22, 2021

Land Degradation in the Western Ghats: The Case of the Kavalappara Landslide in Kerala, India

Nirmala Vasudevan[1], Kaushik Ramanathan[2] and Syali T.S.[3]

ABSTRACT

Landslides are a menacing problem in India. Each year, there are several landslides in various parts of the country. With extreme weather events occurring more frequently and human activity negatively impacting slope stability, the number of fatal landslides has been increasing over the years. One such landslide was the 2019 Kavalappara landslide in the Western Ghats of Kerala State, India. The landslide resulted in the loss of 59 lives and considerable damage to property. Our observations show that human activity exacerbated slope instability—the conversion of natural vegetation to plantations, step cutting of slopes, construction of soak pits causing water to infiltrate into the slope, construction of homes on natural drainage channels, and improper methods of drainage such as directly releasing water from homes into the slope, have all contributed to slope instability. Extremely heavy and continuous rainfall on the days preceding the slide and toe incision of the slope by an inundated tributary of the Chaliyar River proved to be fatal, resulting in a huge debris flow. We recommend some immediate remedial measures and also discuss possible long-term solutions to the problem of slope instability in this region.

Keywords: Landslide, Kavalappara, Western Ghats, Malappuram, Kerala, India.

[1] *Corresponding author.*
Amrita Center for Wireless Networks & Applications, Amrita School of Engineering, Amritapuri, Amrita Vishwa Vidyapeetham, India.
Department of Physics, Amrita School of Arts & Sciences, Amritapuri, Amrita Vishwa Vidyapeetham, India. *E-mail:* nirmalav@am.amrita.edu
[2] Department of Civil Engineering, Amrita School of Engineering, Coimbatore, Amrita Vishwa Vidyapeetham, India. *E-mail:* rkaushik@am.amrita.edu
[3] *Presenting author.*
Department of Physics, Amrita School of Arts & Sciences, Amritapuri, Amrita Vishwa Vidyapeetham, India. *E-mail:* syalitsyamkumar@gmail.com

F-EIR Conference 2021 on
Environment Concerns and its Remediation (ECR21)
Chandigarh, India, October 18–22, 2021

Pulmonary Diseases Due to Bad Indoor Air Quality: Filling the Health Gap in Energy Performance Certificates

Alexandre Soares dos Reis[1], Marta Ferreira Dias[2] and Alice Tavares[3]

ABSTRACT

Good indoor air quality (IAQ) levels in buildings are among the essential benefits and drivers as they lead to better health and comfort of the occupants. However, this research identified a health gap in dwellings' energy performance certificates (EPCs) in Portugal, as IAQ seems not to be appropriately covered. Volatile organic compounds (VOCs) are gases containing various chemicals emitted from liquids or solids. Additionally, biomass-burning stoves are significant contributors to fine particle matter ($PM_{2.5}$) concentrations that may cause cancer and respiratory diseases. Therefore, it is crucial to formulate strategies to control and enhance IAQ. As air pollutants often enter the human body through inhalation, the respiratory system is regularly the main target of Indoor Air Pollution (IAP), resulting in pulmonary diseases and allergies. These facts emphasize the need to track IAQ properly. Depending on indoor air pollutants, several rules and criteria are the basis of the current published work on IAQ indicators. According to our findings, in the planning stage, understandable and straightforward criteria for VOCs, $PM_{2.5}$, and proper ventilation schemes, could help architects and engineers to enhance IAQ. Finally, next-generation EPCs could consider the proposed IAQ score to fill the identified health gap.

Keywords: Energy Performance Certificates, Indoor Air Quality, VOCs, $PM_{2.5}$, Ventilation.

[1] *Presenting & corresponding author.*
Research Unit on Governance, Competitiveness and Public Policies (GOVCOPP), Department of Economics, Management, Industrial Engineering and Tourism (DEGEIT), University of Aveiro, Campus Universitário de Santiago, Aveiro 3810193, Portugal. *E-mail:* alexandre.soares.reis@ua.pt
[2] Research Unit on Governance, Competitiveness and Public Policies (GOVCOPP), Department of Economics, Management, Industrial Engineering and Tourism (DEGEIT), University of Aveiro, Campus Universitário de Santiago, Aveiro 3810193, Portugal.
[3] Research Centre for Risks and Sustainability in Construction (RISCO), Department of Civil Engineering (DECivil), University of Aveiro, Campus Universitário de Santiago, Aveiro 3810193, Portugal.
Centre for Research in Ceramics and Composite Materials (CICECO), Department of Civil Engineering of Univesity of Aveiro, Campus Universitário de Santiago, Aveiro 3810193, Portugal.

F-EIR Conference 2021 on
Environment Concerns and its Remediation (ECR21)
Chandigarh, India, October 18–22, 2021

Self-Consolidating Concrete Produced with Fine CDW

Stéphanie Rocha[1], Joice Soares[2], Thâwanne Costa[2] and Lino Maia[3]

ABSTRACT

Self-consolidating concrete is a concrete technology that seeks to simultaneously achieve fluidity, passing ability and segregation. In its manufacture there is the possibility of using alternative materials, such as ground ceramic, pre-dominantly in construction and demolition waste. Therefore, this study aims to analyze the feasibility of this concrete regarding its properties in the fresh and hardened state when there is replacement of fine sand by ground ceramic aggregate. For this, the flowing tests, flow time, visual stability index, J ring, L box, V funnel and the molding of specimens in three different compositions were carried out, one being the reference compositions and the other two with replacements of 25% and 50% in the natural fine sand by the ceramic waste sand. The mixes in which the replacements were carried out presented good fluidity, without of segregation, however, the passing ability of the mix with 50% of replacement did not fit within the recommended range.

Keywords: Workability, D-flow, Superplasticizer, Segregation.

[1]*Presenting & corresponding author.*
CONSTRUCT-LABEST, Faculty of Engineering (FEUP), University of Porto, RuaDr. Roberto Frias, 4200-465 Porto, Portugal.
FASA, Santo Agostinho Faculty, Av. Osmane Barbosa, 1179-1199 - Jk, Montes Claros-MG, Brazil.
ORCID: 0000-0002-0984-4897. *E-mail:* up202010607@g.uporto.pt
[2]FASA, Santo Agostinho Faculty, Av. Osmane Barbosa, 1179-1199 - Jk, Montes Claros-MG, Brazil.
[3]CONSTRUCT-LABEST, Faculty of Engineering (FEUP), University of Porto, RuaDr. Roberto Frias, 4200-465 Porto, Portugal.
Faculty of Exact Sciences and Engineering, University of Madeira, Campus da Pen-teada, 9020-105 Funchal, Portugal. ORCID: 0000-0002-6371-0179. *E-mail:* linomaia@fe.up.pt

F-EIR Conference 2021 on
Environment Concerns and its Remediation (ECR21)
Chandigarh, India, October 18–22, 2021

UHMWPE/OPA Composite Coatings on Ti6Al4V Alloy as Protective Barriers in a Biological-Like Medium

K. Anaya-Garza[1], M.A. Domínguez-Crespo[2], A.M. Torres-Huerta[3], S.B. Brachetti-Sibaja[4] and J. Moreno Palmerin[5]

ABSTRACT

In this work, UHMWPE/OPA films were coated onto Ti6Al4V alloy substrates using the dip coating method to evaluate their performance as protective barriers against the corrosion process in Simulated Body Solutions. The synthesis parameters to improve the adhesion and electrochemical properties in a Phosphate-buffered saline biological medium were optimized. Structural (XRD) and electrochemical characterizations were also realized to the UHMWPE/OPA coatings. Pure coatings on the metallic substrates acted as protective barriers delaying the interaction between the charged particles of the medium with the surface of the Ti6Al4V alloy. The estimation of the protection efficiency of the UHWMPE films as protective barriers against the corrosion process showed efficiencies of 99% (40 s of immersion time), reducing the corrosion rate up to 0.05 μm y^{-1}. Similarly, the OPA films also presented protective barrier properties, decreasing the corrosion rate up to 1.19 μm y^{-1} with an efficiency of 97% (30 h of immersion time). The synergy of UHMWPE/OPA coatings display an adequate adherence with an important increased in the properties against corrosion.

Keywords: Ti6Al4V Substrates, Biomaterials, UHMWPE/OPA Films, Mechanical Properties, Properties against Corrosion.

[1] *Presenting author.*
  CICATA-Altamira, Instituto Politécnico Nacional, Altamira, México.
[2] *Corresponding author.*
  UPIIH, Instituto Politécnico Nacional, Hidalgo, México. *E-mail:* mdominguezc@ipn.mx
[3] *Corresponding author.*
  UPIIH, Instituto Politécnico Nacional, Hidalgo, México. *E-mail:* atohuer@hotmail.com
[4] IT de Ciudad Madero, Tecnológico Nacional de México, Tamps. México.
[5] Universidad de Guanajuato, Guanajuato, México.

F-EIR Conference 2021 on
Environment Concerns and its Remediation (ECR21)
Chandigarh, India, October 18–22, 2021

Carbon Foams Preparation by Replica Method Using Activated Carbon from Sugarcane Bagasse and Orange Peel Wastes: Applications to Metal Ions Removal

A.I. Licona-Aguilar[1], A.M. Torres-Huerta[2], M.A. Domínguez-Crespo[3], E. Conde-Barajas[4], M.L.X. Negrete-Rodríguez[4], D. Palma-Ramírez[5] and A.E. Rodríguez-Salazar[6]

ABSTRACT

In this work, highly efficient removal of heavy metals, from biochar and AC from sugarcane bagasse (SCB) and orange peel (OP), respectively, were obtained. Physical - chemical activation method in acid medium (H_3PO_4, 85 wt.%), followed by an activation at high temperature (500 and 700°C), was performed. These materials were used to produce carbon foams (CF) using the replica method and evaluated by their adsorbent capacity for the removal of heavy metals from synthetic water. The obtained products, from SCB and AC from OP, having surface area of (4–6 m^2/g pore diameter size) and 249 m^2/g, were classified as biochar and AC, respectively. Results indicate that among the pH studied (3, 5 and 8), biochar from SCB activated at 500°C, showed the best efficiency for Pb (94%) and Cu (85%) removal at pH 8. A similar efficiency removal was observed for biochar from OP either at 500°C (Pb:90% and Cu:78%) or 700°C (Pb:93% and Cu:77%) at the same pH, whereas CF from OP 700°C had the better efficiency at pH 5 in comparison to all the systems showing 99% and 97% for Pb and Cu removal showing a minimum concentration remaining of Lead (Pb 0.06 mg/L) and Copper (Cu 0.3 mg/L) considering an initial concentration of 50 mg/L of each metal. The study confirms that agro-industrial wastes can be effectively valorized to produce adsorbents carbon foam for wastewater treatment applications.

Keywords: Agroindustrial Wastes Adsorption, Carbon Foams, Metal Removal, Wastewater Treatment.

[1] CICATA-Altamira, Instituto Politécnico Nacional, Altamira, México.

[2] *Corresponding author.* UPIIH, Instituto Politécnico Nacional, Hidalgo, México. *E-mail:* atohuer@hotmail.com

[3] *Corresponding author.* UPIIH, Instituto Politécnico Nacional, Hidalgo, México. *E-mail:* mdominguezc@ipn.mx

[4] Laboratory of Environmental Biotechnology, Department environmental engineering, Instituto Tecnológico de Celaya, Guanajuato, México.

[5] CMPL, Instituto Politécnico Nacional, México City, México.

[6] CICATA Querétaro, Instituto Politécnico Nacional, Querétaro. México.

Evaluating the Application of Metaheuristic Approaches for Flood Simulation Using GIS: Baitarani River Basin, India

Sandeep Samantaray[1], Sambit Sawan Das[2], Abinash Sahoo[3]
and Deba Prakash Satapathy[4]

ABSTRACT

Flood is one of the significant natural disasters, which is treated as one of the main global concerns which increased occurrence has led to an increase in mortality rates and economic losses. Various methods have been developed and proposed for the analysis of this natural disaster. In recent years, data mining approaches such as artificial neural network (ANN) techniques are being increasingly used for flood modeling. The specific objective of this study is to develop a flood model using various flood causative factors using ANN, adaptive neuro-fuzzy Inference system (ANFIS) and hybrid ANFIS-GOA (grasshopper optimization algorithm) techniques with the help of geographic information system (GIS) and simulate flood-prone areas affected by Baitarani River basin flowing mostly through Keonjhar district in northern part of Odisha, India. Thirteen factors, namely slope, aspect, altitude, curvature, geology, soil, landuse/ cover (LULC), topographic wetness index (TWI), stream power index (SPI), terrain roughness index (TRI), sediment transport index (STI) and distance from rivers and roads, were utilized. In the context of objective weight assignments, the ANN is used to directly produce water levels and then the flood map is constructed in GIS. To measure the performance of the model, three criteria performances, including a coefficient of determination (R^2), Nash–Sutcliffe Efficiency (NSE), and the root mean square error (RMSE) are used. The verification results showed satisfactory agreement between the predicted and the real hydrological records. The validation results revealed that the hybrid ANFIS-GOA model

[1] Assistant Professor, Department of Civil Engineering, College of Engineering and Technology, India.
   E-mail: sandeep1139_rs@civil.nits.ac.in
[2] *Presenting & corresponding author.*
   M.Tech Research Scholar, Department of Civil Engineering, College of Engineering and Technology, India.
   E-mail: sambitsawan@gmail.com
[3] Ph.D. Research Scholar, Department of Civil Engineering, NIT Silchar, India.
   E-mail: bablusahoo1992@gmail.com
[4] Associate Professor, Department of Civil Engineering, College of Engineering and Technology, India.
   E-mail: dpsatapathy@cet.edu.in

(with an accuracy of 99.20%) was better at predicting flood than ANFIS and ANN models with an accuracy of 92.62% and 89.56% respectively. The results of this study could be used to help local and national government plan for the future and develop appropriate (to the local environmental conditions) new infrastructure to protect the lives and property of the people of Keonjhar. Based on the findings, we were able to derive a more effective method for analyzing flood susceptibility.

Keywords: Flood, Baitarani River, ANFIS-GOA, ANN.

F-EIR Conference 2021 on
Environment Concerns and its Remediation (ECR21)
Chandigarh, India, October 18–22, 2021

Improving Accuracy of SVM for Monthly Sediment Load Perdition Using Harris Hawks Optimization

Sandeep Samantaray[1], Abinash Sahoo[2] and Deba Prakash Satapathy[3]

ABSTRACT

The suspended sediment load (SSL) prediction is one of the most important issues in water engineering. Accurate prediction of suspended sediment (SS) concentration is a difficult task for water resource projects. In recent years, methodologies such as artificial intelligence (AI) algorithms have been applied for sediment load estimation and these models have provided efficient results. In this article, the support vector machine (SVM) were used to estimate SSL at two stations of Baitarani river basin, Odisha, India. A new evolutionary algorithm, namely, the Harris hawks optimization (HHO), was used to enhance the precision of the SVM models for predicting monthly SSL. The lagged rainfall, temperature, stage, runoff were used as the inputs to the models. In this research, the abilities of the SVM-HHO were benchmarked with SVM-particle swarm optimization (PSO), and SVM-GWO (Grey wolves optimization). Data were separated into two subsets (training, and testing) and SSL was predicted where the reliability of utilized approaches was assessed by statistical criterion including the coefficient of determination (R^2), mean absolute percentage error (MAPE), root mean square error (RMSE), Nash-Sutcliffe efficiency (NSE) and PBAIS. The NSE of SVM-HHO was 0.993 higher than those of SVM-GWO, SVM-PSO, and SVM models in the training level for Baitarani River. It was concluded that the SVM-HHO had the highest value of MAPE, R^2 and minimum of RMSE among other models in the Baitarani River. This finding reveals the superiority of HHO meta-heuristic algorithm in improving the accuracy of the standard AI in forecasting the monthly SSL. The performance of the HHO algorithm for escaping local optima makes SVM-HHO as a powerful tool for estimating sediment load. In addition, for cases studies, there is a need to develop real-time forecasting model, and hence, the time consuming for the model execution should be lessened.

Keywords: SVM-HHO, SVM-PSO, SVM-GWO, Baitarani River, Sediment Load.

[1] Ph.D Research Scholar, Department of Civil Engineering, NIT Silchar, India.
  E-mail: sandeep1139_rs@civil.nits.ac.in
[2] *Presenting & corresponding author.*
  Ph.D Research Scholar, Department of Civil Engineering, NIT Silchar, India.
  E-mail: bablusahoo1992@gmail.com
[3] Associate Professor, Department of Civil Engineering, College of Engineering and Technology, India.
  E-mail: dpsatapathy@cet.edu.in

Graphical Abstract

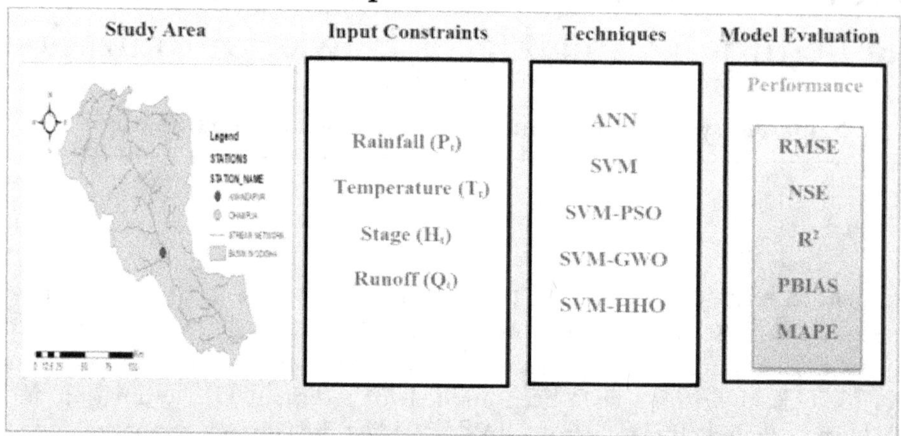

F-EIR Conference 2021 on
Environment Concerns and its Remediation (ECR21)
Chandigarh, India, October 18–22, 2021

Impact of Waste Ceramic Tiles as Partial Replacement of Fine and Coarse Aggregate in Concrete

A. Sivakumar[1], S. Srividhya[2], V. Sathiyamoorthy[3], M. Seenivasan[4] and M.R. Subbarayan[5]

ABSTRACT

A substantial amount of ceramic waste is generated in the form of damaged or broken ceramic products. Minimizing solid waste disposal is one of the potential solutions to a cleaner environment. Because of its interesting applications and low cost, the use of solid wastes in sustainable construction has received a lot of attention. The employment of waste ceramic as a fine and coarse aggregate replacement was studied in this current research. Specimens with ceramic waste as fine and coarse aggregates at varying percentage levels of 0%, 10%, 20%, 30%, 40%, and 50% was prepared for the current study. The influence of ceramic fine and coarse aggregate on the characteristics of concrete was investigated using fresh, mechanical, and durability properties. Based on the findings, the optimal aggregate replacement level ratio was found to be 20%. When optimized quantity of 20% ceramic waste was used in concrete, the compressive strength, split tensile strength, and flexural strength were higher than that of the other mixtures. The strength of the concrete was lower with higher quantities of ceramic waste replacement than with the conventional mix.

Keywords: Ceramic Waste, Bond Strength, Aggregate, Sustainable Concrete, Compressive Strength.

[1] *Presenting & corresponding author.*
Department of Mechanical Engineering, Varuvan Vadivelan Institute of Technology, Dharmapuri, India.
E-mail: sirarira@gmail.com
[2] Presenting & corresponding author.
Department of Civil Engineering, Varuvan Vadivelan Institute of Technology, Dharmapuri, India.
E-mail: srividhya132@gmail.com
[3] Department of Mechanical Engineering, KSRM College of Engineering, Kadapa, India.
E-mail: drvsathiyamoorthy@gmai.com
[4] Department of Mechanical Engineering, KGiSL Institute of Technology, Coimbatore India.
E-mail: seenivasan.m@kgkite.ac.in
[5] Department of Mechanical Engineering, Jayam College of Engineering and Technology, Dharmapuri, India.
E-mail: subbarayan7779@gmail.com

F-EIR Conference 2021 on
Environment Concerns and its Remediation (ECR21)
Chandigarh, India, October 18–22, 2021

Study of Sustainability Aspects of Chambal River Linking System Using Reservoir Simulation Techniques

Rajesh Kumar Jain[1], Rajesh Goyal[2] and Dulal Goldar[3]

ABSTRACT

The present problem is confined to the Sustainability of four reservoirs of Chambal catchment in Madhya Pradesh and Rajasthan. These are Mohanpura on River Newaj, Kundaliya on River Kalisindh and Gandhi Sagar & Rana Pratap Sagar on River Chambal. The first case is when reservoirs operated without linking (case1) which shows that water is available for transfer from proposed Mohanpura and Kundaliya reservoirs for diversion and when a diversion from Mohanpura reservoir joins the Kundaliya reservoir a sufficient amount of water can be transferred to downstream to either Gandhi Sagar (2a) or Rana Pratap Sagar (2b). The analysis of case 2(a) and 2(b) also shows that for existing proposed diversions the proposed reservoir capacities of Mohanpura and Kundaliya reservoirs can be reduced substantially and that the existing proposed diversions have a positive impact on the hydropower generation at Gandhi Sagar reservoir.

Keywords: *Reservoir, Sustainibility, Mohanpura, Kundaliya, Gandhi Sagar, Rana Pratap Sagar.*

[1] *Presenting & corresponding author.*
Research Scholar, NICMAR Delhi Lingaya Vidyapeeth, Faridabad, India.
E-mail: rkjainnwda@gmail.com
[2] Rajesh Goyal Dean and Professor NICMAR, Bahadurgarh, India.
[3] Professor, Lingaya Vidyapeeth, Faridabad, India.

F-EIR Conference 2021 on
Environment Concerns and its Remediation (ECR21)
Chandigarh, India, October 18–22, 2021

Effective Utilization of Waste Eggshell Powder in Cement Mortar

K. Nandhini[1] and J. Karthikeyan[2]

ABSTRACT

Solid waste management is one of the challenging issues faced by the developing countries. The developing countries find it very critical to manage the disposal of waste generation. India ranks 3rd in the production of egg, generating about 3.8 billion kilograms annually. This growth in the production of larger rate is mainly due to the hike in domestic consumption. This could lead to larger generation of solid waste. To overcome these issues, eggshell could be effectively incorporated into concrete production as cementitious blends. This paper investigates the suitability of eggshell powder as partial substitute to cement. The material characterization techniques like SEM and XRD were used to investigate the feasibility of using eggshell powder. The study was conducted by producing cement mortars in the ratio of 1:3:0.5. Then the cement was partially replaced by eggshell powder at 5%, 10%, 15%, 20% by its weight. The fresh property was assessed using the flow table test, while the hardened property was determined using the compressive strength of the cement mortar at 7, 14, and, 28 days. From the test results, 10% eggshell powder when utilized in mortar gives optimum compressive strength.

Keywords: Eggshell Powder, Sustainability, Material Characterization, Slump Flow, Compressive Strength.

Graphical Abstract

---

[1] *Presenting & corresponding author.*
Post-Doctoral Fellow, Department of Civil Engineering, National Institute of Technology Tiruchirappalli, India. *E-mail:* knandhini@nitt.edu
[2] Associate Professor, Department of Civil Engineering, National Institute of Technology Tiruchirappalli, India. *E-mail:* jk@nitt.edu

F-EIR Conference 2021 on
Environment Concerns and its Remediation (ECR21)
Chandigarh, India, October 18–22, 2021

Mechanical Properties of Concrete with Seashell Waste as Partial Replacement of Cement and Aggregate

P. Sangeetha[1], M. Shanmugapriya[2], K. Santhosh Saravanan[3],
P. Prabhakaran[4] and V. Shashankar[5]

ABSTRACT

Seashell is a very hard and protective outer layer produced by an animal that lives in the sea. Empty seashells are mostly found in the shore and collected as waste by beachcombers. This marine by-product can be used to partial replacement of coarse aggregate and cement in concrete. This paper describes the potential use of seashell powder and aggregate as cement and aggregate replacement for making a greener concrete. The study focused on the effect of seashell powder and aggregate in concrete with respect to compressive, split tensile and flexural strength after 7, 28, 56 and 90 days of curing of concrete specimen. The replacement of cement by seashell powder are 5%, 10% and 15% and replacement of coarse aggregate by seashell aggregate are 10%, 20% and 30%. The properties of seashell concrete were compared with control mix specimen of M25 grade of concrete. It was found that replacement of cement and coarse aggregate by seashell products enhances the strength significantly and enable proper utilisation of these seashell waste as sustainable material for the concrete.

Keywords: *Seashell Powder, Seashell Coarse Aggregate, Compressive Strength, Flexural Strength, Tensile Strength.*

[1] *Corresponding author.*
Associate Professor, Sri Sivasubramaniya Nadar College of Engineering, Chennai 603110, India.
E-mail: sangeethap@ssn.edu.in

[2] Assistant Professor, Department of Mathematics, Sri Sivasubramaniya Nadar College of Engineering, Chennai 603110, India. *E-mail:* shanmugapriyam@ssn.edu.in

[3] *Presenting author.*
UG Student, Department of Civil Engineering, Sri Sivasubramaniya Nadar College of Engineering, Chennai 603110, India. *E-mail:* santhoshsaravanan047@civil.ssn.edu.in

[4] UG Student, Department of Civil Engineering, Sri Sivasubramaniya Nadar College of Engineering, Chennai 603110, India. *E-mail:* prabha07lpm@gmail.com

[5] UG Student, Department of Civil Engineering, Sri Sivasubramaniya Nadar College of Engineering, Chennai 603110, India. *E-mail:* shashankar049@civil.ssn.edu.in

F-EIR Conference 2021 on
Environment Concerns and its Remediation (ECR21)
Chandigarh, India, October 18–22, 2021

Review on Application of Membrane Technology to Control Environmental Pollution

Praful G. Bansod[1], Jaykumar Bhasarkar[2], Swapnil Dharaskar[3] and Shyam Kodape[4]

ABSTRACT

The available natural energy sources are insufficient to meet everyone's needs. Due to an increased population, dwindling natural resources, limited natural sources, and environmental issues, it is critical to focus on reproducing, recycling, and reused techniques. Coal is a major source of energy generation in India, but it causes significant environmental pollution. The use of petroleum oil for transportation contributes to global warming, and domestic and industrial waste contributes to pollution. It is critical to consider waste water treatment as well as alternative energy sources such as biofuel. Conventional technology is insufficient in terms of both economics and the environment. Membrane technology is an effective and efficient technology. This paper provides an overview of recent research on membrane technology applications in waste water treatment, biofuel production (such as biodiesel and bioethanol). It also discusses membrane reactors, membrane bioreactors, and mixed matrix membrane reactor applications for biodiesel production, as well as hybrid pervaporization membrane bioreactors, integrated pervaporization membrane bioreactors for bioethanol production, the membrane bioreactors application for waste water treatment

Keywords: *Membrane, Waste Treatment, Biodiesel, Bioethanol, Waste Water Treatment.*

[1] *Presenting and corresponding author.*
Department of Chemical Engineering, Sinhgad College of Engineering, Pune 411041, India.
E-mail: praful1vaisu@gmail.com
[2] Department of Pulp and Paper Technology, Laxminarayan Institute of Technology, R.T.M. Nagpur University, Nagpur, India. *E-mail:* Jaykumar.ppt@gmail.com
[3] Department of Chemical Engineering, Pandit Deendayal Energy University, Gandhinagar 382426, India.
E-mail: swapnildharaskar11@gmail.com
[4] Department of Chemical Engineering, Visvesvaraya National Institute of Technology, Nagpur 440010, India.
E-mail: samkodape@gmail.com

F-EIR Conference 2021 on
Environment Concerns and its Remediation (ECR21)
Chandigarh, India, October 18–22, 2021

Effect of Various Surfactant Templates on the Physicochemical Properties of $CoFe_2O_4$ Nanoparticles

Anirban Dey[1] and Bharti Saini[2]

ABSTRACT

Transition metal oxide ferrite nanoparticles exhibit some unique physical and chemical properties which are significantly different from the bulk materials due to its extremely small size and high surface area. Among the various ferrite nanoparticles inverse spinel ferrites such as $CoFe_2O_4$ posseses excellent chemical stability, high coercivity, moderate saturation magnetization as well as high crystalline anisotropy. These properties make it a promising agent for drug delivery, catalysis and magnetic storage devices. The addition of surfactant in the synthesis process of ferrite magnetic nanoparticles facilitate in controlling the particle growth, and agglomeration between the magnetic nanoparticles because of the steric hindrance and stabilization properties of surfactant In this study a series of $CoFe_2O_4$ nanoparticles have been synthesized successfully in presence of various surfactant viz. cationic, anionic and non-ionic. $CoFe_2O_4$ nanoparticles was prepared in aqueous medium by Co-precipitation technique using Cobalt nitrate (Co^{2+}) and Ferric nitrate (Fe^{3+}) as the precursor and CTAB, SDS and Tween-20 were used as the templating agent. The Precipitation was achieved by the rapid addition of a strong base, NaOH. The structural as well as the magnetic properties were investigated by XRD, FT-IR, FESEM and VSM analysis. The XRD analysis of the nanoparticles confirms that the cubic spinel phase of the $CoFe_2O_4$ nanoparticles was retained after the addition of templating agent.

Keywords: $CoFe_2O_4$ Nanoparticle, Co-precipitation, Surafctant, Templating Agent.

---

[1] *Presenting & corresponding author.*
Department of Chemical Engineering, Pandit Deendayal Energy University, Gandhinagar 382007, India.
E-mail: anirban.dey@sot.pdpu.ac.in
[2] Department of Chemical Engineering, Pandit Deendayal Energy University, Gandhinagar 382007, India.

Experimental Study on Eccentrically Loaded Rectangular Footing on Sand for Embedded Condition to Enhance Serviceability of Structures

Gaurav Bharti[1], Bishnu Kant Shukla[2], Varadharajan Srinivasan[3] and Vaishnavi Bansal[4]

ABSTRACT

After the introduction of Terzaghi's theory in the field of geotechnical engineering regarding ultimate bearing capacity of shallow foundation, work has also been conducted in both experimental and theoretical in this field. But most of the studies were related to only vertically apply central loading on the footing. It is important to know durability of structure and the effect of various types of loading on bearing capacity of footing. At that point of time there was no study conducted regarding inclined loading and eccentrically loading on the footing. But later came the theory of Meyerhof in 1953 who suggested an empirical procedure regarding the estimation of ultimate (maximum) bearing capacity of shallow foundation for the eccentric loading condition. In the present experimental work, the sample tests have been computed to evaluate the maximum bearing capacity of shallow foundation for embedded condition in rectangular footing. Various test is carried out on sample footing of size 18×24cm. The tests is carried out on dense sand. The relative density (Dr) of dense sand sustained during the test is 62.780%. The eccentricity ratio is varied from 0 to 0.166 and depth is varied from 4.5 cm to 18cm starting from 4.5 cm (0.25B). Maximum capacity of bearing has been computed for eccentric loading condition and enhancement in structural models was determined.

Keywords: Bearing Capacity, Durability, Embedded, Eccentric Loading, Footing, Serviceability Model.

[1] *Presenting & corresponding author.*
Department of Civil Engineering, National Institute of technology, Kurukshetra, India.
Department of Civil Engineering, JSS Academy of Technical Education, Noida, India.
E-mail: gauravbharti000@gmail.com
[2] *Corresponding author.*
Department of Civil Engineering, JSS Academy of Technical Education, Noida, India.
E-mail: bishnukantshukla@gmail.com
[3] Department of Civil Engineering, JSS Academy of Technical Education, Noida, India.
E-mail: svrajan82@gmail.com
[4] Department of Civil Engineering, JSS Academy of Technical Education, Noida, India.
E-mail: gvaishnavibansal@gmail.com

Assessing Outdoor Thermal Comfort at an Urban Park during Summer in the Hot Semi-arid Region of India

Pardeep Kumar[1] and Amit Sharma[2]

ABSTRACT

Urban parks play an essential role in urban settings; significantly contribute to the health of every age group person. Parks provide opportunities for families to connect with nature and breathe in the fresh air. Due to global climate change and increased urbanization in the past few decades, extreme heat can be experienced in urban areas. Consequently, people opt to stay most of the time indoors and live a sedentary lifestyle. Mental and physical health issues arise primarily due to a sedentary lifestyle. Staying at parks for a longer duration could promote stress reduction and perceived physical health. The present study aims to assess the thermal comfort conditions at an urban park in the hot semi-arid region of Haryana, India. The present study investigated the outdoor thermal comfort range and thermal sensations of visitors at a park during the summer season using the onsite monitoring of the microclimate parameters and questionnaire survey in the hot-semi arid region of India. Thermal comfort indices, Physiological equivalent temperature (PET) and Universal Thermal Climate Index (UTCI) have been applied to investigate the outdoor thermal comfort conditions. The ASHRAE seven-point sensation scale has been used to record the visitors' thermal sensations. The results indicated that: 1) UTCI was found to be the most suitable index to investigate the OTC conditions. The neutral UTCI and PET ranged within 28.03°C to 35.61°C and 24.04°C to 37.55°C, respectively. 2) The neutral PET and UTCI were found to be 30.79°C and 31.82°C, respectively.3) Dry bulb temperature is the most significant thermal comfort parameter affecting visitors' thermal sensations, followed by mean radiant temperature.4) PET and UTCI found to be most significantly affected by globe temperature. The study's outcome could provide theoretical design reference to urban designers to develop new parks and existing parks, ultimately promoting public health.

Keywords: Outdoor Thermal Comfort, Urban Parks, PET, UTCI, Neutral Temperature Range.

[1] *Presenting & corresponding author.*
Doctoral Candidate, Department of Mechanical Engineering, Deenbandhu Chhotu Ram University of Science and Technology, Sonipat, India. *E-mail:* kumar.pardeepmech@gmail.com
[2] *Corresponding author.*
Assistant Professor, Department of Mechanical Engineering, Deenbandhu Chhotu Ram University of Science and Technology, Sonipat, India. *E-mail:* amitsharma.me@dcrustm.org

F-EIR Conference 2021 on
Environment Concerns and its Remediation (ECR21)
Chandigarh, India, October 18–22, 2021

Improvement in Behaviour of Black Cotton Soil using Copper Slag

Gaurav Bharti[1], **Shubham Jagudi**[2] **and Bishnu Kant Shukla**[3]

ABSTRACT

The present experimental study aimed at determining the behaviour of black cotton soil reinforced with Copper slag in a random manner. Use of any waste materials like slag (copper) in soil stabilization is both environmental and engineering favourable. The soil used is a type of expansive soil obtained from District, Visakhapatnam, Andhra Pradesh, India. The soil is obtained by removing the surface layer and collected from the depth 0.50 m from the natural ground level. The samples of expansive soil were prepared at four different rate percentages share of copper slag (5.0%, 10.0%, 15.0% and 20.0% by mass of soil). Unconfined compression test is carried out for soil sample after 0, 7 and 14 day curing periods. CBR test for 4 days soaking period for soil samples is also carried out. Modified proctor test is also computed for present evaluation.

Keywords: Black Cotton Soil, Soil Stabilization, Copper Slag, Environment, CBR Test, Sustainability.

[1] *Presenting author.*
Department of Civil Engineering Academy of Technical Education, Noida, India. Department of Civil Engineering, National Institute of Technology, Kurukshetra, India. *E-mail:* gauravbharti000@gmail.com
[2] Department of Civil Engineering, National Institute of Technology, Kurukshetra, India.
E-mail: shubham543jagudi@gmail.com
[3] *Corresponding author.*
Department of Civil Engineering Academy of Technical Education, Noida, India.
E-mail: bishnukantshukla@gmail.com

Synthesis and Characterization of Copolymer Adsorbent for Crystal Violet Dye Removal from Water

Bharti Saini[1] and Anirban Dey[2]

ABSTRACT

Existence of a very lesser quantity of dyes in water, which are still highly visible, extremely affects the quality and clearness of water bodies such as lakes, rivers, ponds and others water bodies, leading to harm to the aquatic environment. So it is significant to eliminate the harmful dyes which are very hazardous to the living organisms. Treatment of wastewater is challenging because dyes are water soluble so its removal becomes tough. In this work synthetic copolymer adsorbent of AA and AMP was successfully synthesized at many compositions. All the characteristic absorption bands of PAA and PAMP are observed by FTIR analysis, which indicates the successful copolymerization of AA and AMP. Adsorption experiments were accomplished at different contact time, adsorbent dose, pH conditions and temperature. 0.1 g dose of copolymer adsorbent has been found to be optimum for adsorption study. The kinetic study of adsorption was also investigated using pseudo first order kinetic model. From the study, it was establish that the prepared copolymer adsorbent was proficient to eliminate crystal violet dye at various temperature and time. Maximum 98% of crystal violet dye removal was achieved at high pH around 10. A maximum adsorption capacity of 9.8 mg/g was obtained for copolymer adsorbent of 1:1 ratio. Copolymer adsorbent (AA-AMP) provide a novel and effective type of adsorbents to remove CV dye from the aqueous solution.

Keywords: Copolymer, Adsorbent, Adsorption, Crystal Violet, Dye Wastewater Treatment.

[1] *Presenting & corresponding author.*
Department of Chemical Engineering, School of Technology, Pandit Deendayal Energy University, Gandhinagar, India. *E-mail:* bharti.saini@sot.pdpu.ac.in
[2] Department of Chemical Engineering, School of Technology, Pandit Deendayal Energy University, Gandhinagar, India. *E-mail:* anirban.dey@sot.pdpu.ac.in

F-EIR Conference 2021 on
Environment Concerns and its Remediation (ECR21)
Chandigarh, India, October 18–22, 2021

Developing Multiple Natural Dyes from Flowers of *Delonix Elata*: Colourimetric Analysis of Dyed Yarns and Antimicrobial Evaluation

R. Praveena[1] and C. Kavitha[2]

ABSTRACT

In today's era, need for new sources of dyes is a pre-requisite factor for sustainable development in the field of textiles. Exploration and introduction of novel natural dye source is mandatory for re-introduction of plant colorant based dyes into modern textile dyeing operations. In the present study, yellow flavonoid dyes obtained from *Delonix elata* flowers were dyed on cotton and silk using traditional techniques. Further, dyeing experiment was carried out using combination of natural and metal salt mordants. It is observed that premordanting with myrobalan followed by simultaneous mordanting with metal salts such as Alum, Stannous chloride, Copper sulphate, Zinc chloride, Ferrous sulphate , Aluminium acetate enhances the dyeing ability. Naturally dyed silk fibers yielded better shades varying from yellow to brown than cotton which is evident from the Lab color space coordinates and colour strength values. Fastness properties ranged from good to excellent, while light fastness was fair to good. Further, the antimicrobial properties of the finished cotton and silk substrates was evaluated by American Association of Textile Chemists and Colorists test method 100–2004 using two human pathogenic strains, *Staphylococcus aureus* and *Klebsiella pneumonia* and was found to exhibit excellent activity.

Keywords: *D. Elata, Mordant, Coloumetric Values, Fastness, Natural Dyeing, Antimicrobial Activity.*

[1] *Presenting & corresponding author.*
Assistant Professor(s), Department of Chemistry, Bannari Amman Institute of Technology, Sathyamangalam, India. ***E-mail:*** praveenar@bitsathy.ac.in
[2] Assistant Professor(s), Department of Chemistry, Bannari Amman Institute of Technology, Sathyamangalam, India. ***E-mail:*** kavithac@bitsathy.ac.in

F-EIR Conference 2021 on
Environment Concerns and its Remediation (ECR21)
Chandigarh, India, October 18–22, 2021

Removal of Nickel (II) Ions from Polluted Aqueous System Using Naturally Ocurring Bio Sorbent

C. Kavitha[1] and R. Praveena[2]

Abstract

Nickel (II) ions are among the most harmful pollutants in several industrial discharges being discharged into the environment as industrial wastes, causing serious soil and water pollution. Nickel can exist in several valence states. In aquatic stream, nickel is probably complexed with high affinity for soil by organic matter giving soluble and insoluble compounds, which are having ill effects on living organisms. Most probably the part which is not complexed with organic matter exists in soil solutions predominantly as divalent cations. The aim of the present study is to investigate the removal of nickel ions from aqueous solutions using cotton seeds. Batch experimental studies were conducted to evaluate the optimum conditions for Ni(II) removal by changing relevant parameters such as initial pH of solution and dosage of adsorbent used. The absorbance measurements were carried out on a Systronics UV-Visible spectrophotometer with 1 cm matched quartz cells. The pH of buffer solutions was monitored by using Systronic digital pH meter (India). From the absorbance measurements, the optimum pH for removal of Ni(II) ions is between 4– 7 at the dosage of 50 mg of cotton seed adsorbent allowed at an equilibration time of 5 hours. At pH 7, the amount of adsorbent is optimized and found to be 50 mg/100 ml of Ni (II) solution for effective removal upto 90%. Also, based on the experimental results, it can be concluded that cotton seed has the greater potential of application as an efficient adsorbent for the removal of heavy metals from aqueous solutions. Also the extent of removal can be effectively studied using spectrometer.

Keywords: *Nickel (II), Cotton Seeds, Absorbents, Spectrophotometer.*

[1] *Presenting & corresponding author.*
Department of Chemistry, Bannari Amman Institute of Technology, Erode, India.
E-mail: ckaviuvaraja@gmail.com
[2] Department of Chemistry, Bannari Amman Institute of Technology, Erode, India.
E-mail: praveethang@gmail.com

F-EIR Conference 2021 on
Environment Concerns and its Remediation (ECR21)
Chandigarh, India, October 18–22, 2021

Review on Bacteria based Cementitious Matrix for Sustainable Building Construction

Krishna Kumar Maurya[1], Anupam Rawat[2] and Rama Shanker[3]

ABSTRACT

Sustainable construction is obligatory for the development of the infrastructures to facilitate structural integrity. Incorporation of bacteria in cementitious materials is likewise a technique for sustainable concrete construction development and structural strengthening. The inclusion of bacteria in the cementitious-based matrix is prominent for microbial induced calcite precipitation through the respiratory metabolism process, environmentally friendly biodegradable material. This review paper is focused on parameters that are significant to develop bacterial concrete. In this paper, the bacterial concrete phenomenon and its effect on concrete have been presented. Further, the consequences of bacteria types, application methods, growth media for bacteria, and optimal cell concentration were deliberated. The advantages and disadvantages due to different kinds of bacteria on concrete have been discussed. The incorporation methods of bacteria viz., direct and indirect were enlightened, further their suitability was analyzed. Moreover, the effects of bacteria application on the mechanical properties of concrete were examined. Crack healing phenomenon in concrete due to microbial-induced calcite precipitation has been elaborated. The reviewed article will result in innovative research directions to the civil industry and valuable for environment friendly sustainable concrete construction of real-life infrastructures.

Keywords: *Bacterial Concrete, Bacteria Type, Incorporation Methods, Optimum Cell Concentration, Crack Healing, Culture Media.*

[1] *Presenting & corresponding author.*
Ph.D. Research Scholar, Department of Civil Engineering, Motilal Nehru National Institute of Technology Allahabad, Prayagraj, 211004, India. *E-mail:* krish01990@gmail.com
[2] Assistant Professor, Department of Civil Engineering, Motilal Nehru National Institute of Technology Allahabad, Prayagraj, 211004, India. *E-mail:* anupam@mnnit.ac.in
[3] Associate Professor, Department of Civil Engineering, Motilal Nehru National Institute of Technology Allahabad, Prayagraj, 211004, India. *E-mail:* ramashanker@mnnit.ac.in

F-EIR Conference 2021 on
Environment Concerns and its Remediation (ECR21)
Chandigarh, India, October 18–22, 2021

An Overview of Highway Failure and Its Maintenance on Old NH-2

Parveen Berwal[1], Maneesh Pal[2], Ishtiaque Ali[3], Shilpa Singla[4] and Rajesh Goyal[5]

ABSTRACT

Full-grown transportation infrastructure is very important for social, industrial, cultural, and economic development. The main focus of this paper shows an examination to track down the adaptable asphalt distress and rigid pavement failure types and recognize the causes and select the best maintenance for the disappointment of flexible asphalt along the 10 km distance of old NH-2 (current NH-19) in Fatehpur district in Uttar Pradesh. It includes 8 km flexible pavement and 2 km rigid pavement. There are numerous kinds of distresses that occur in the road like various kinds of breaks, potholes raveling water dying, layering and pushing, misery and rutting. The fundamental driver of flexible asphalt is poor mixing of bituminous, heavy loads, poor drainage conditions, and heavy rainfall. The failures are removed by adequate planning, inspection, and treatment. These failures create problems for motorists, passengers and also increase cost maintenance. National highway authority of India suggests that to try maintenance regularly with the needs of critical support and accessibility of assets. Which is found during this examination. The present study identifies the different types and classifications of faults in flexible pavements and the sources of these faults and how to correct them also the current pavement flaws and associated maintenance procedures.

Keywords: Pot Hole, Raveling, Maintenance, Transportation.

[1] *Corresponding author.*
Civil Engineering Department, Galgotias College of Engineering and Technology, Greater Noida, Uttar Pradesh 201310, India. *E-mail:* parveenberwal@gmail.com
[2] *Presenting author.*
Civil Engineering Department, Galgotias College of Engineering and Technology, Greater Noida, Uttar Pradesh 201310, India.
[3] Civil Engineering Department, Galgotias College of Engineering and Technology, Greater Noida, Uttar Pradesh 201310, India.
[4] Civil Engineering Department, Sant Longowal Institute of Engineering and Technology, Sangrur, Punjab 148106, India.
[5] School of Construction Management, NICMAR Delhi NCR, Bahadurgarh 124507, India.

F-EIR Conference 2021 on
Environment Concerns and its Remediation (ECR21)
Chandigarh, India, October 18–22, 2021

An Overview on Study of Traffic Monitoring System

Parveen Berwal[1], Gopesh Kaushik[2], Kunal Chandra[2], Kirti Upadhyay[3],
Deepak Gupta[2] and Rajesh Goyal[4]

ABSTRACT

In the recent past, the term traffic congestion has emerged as one of the most brutal challenges to face for engineers, governments, planners, and policymakers in almost every country. With ongoing advancements in social and economic structures, fabricated by vehicle-oriented urban development, they have established congestion as an inevitable reality of urban life. Predominantly there are various factors that play a humongous role in aggravating this problem some of them are listed as: (Speed, Travel time, Delay, Volume), etc. Subsequently, every research demands exhaustive data collection and based upon that, reaching upon suitable methods to overcome the widely spread problem. Speed has emerged as the most important traffic measurement to identify congestion. This phenomenon has been widely classified into two types (i) Recurring and (ii) Nonrecurring. According to the United States Department of Transportation Federal Highway Administration- "alone, nonrecurring congestion is responsible for more than 50% of the total traffic congestion, whereas, recurring congestion contributes for about 40%. Extensive collection of data is aimed to be done at several locations in order to measure the maximum number of traffic characteristics. The main emphasis of this paper is aimed towards understanding the recurring urban congestion, its types, and methods to curtail it with the least available manpower. Literature reviews regarding this problem reveal some interesting insights.

Keywords: Congestion, Speed, Travel Time, Recurring.

[1] *Corresponding author.*
Civil Engineering Department, Galgotias College of Engineering and Technology, Greater Noida, Uttar Pradesh 201310, India. *E-mail:* parveenberwal@gmail.com
[2] Civil Engineering Department, Galgotias College of Engineering and Technology, Greater Noida, Uttar Pradesh 201310, India.
[3] *Presenting author.*
Civil Engineering Department, Galgotias College of Engineering and Technology, Greater Noida, Uttar Pradesh 201310, India.
[4] School of Construction Management, NICMAR Delhi NCR, Bahadurgarh, 124507, India.

F-EIR Conference 2021 on
Environment Concerns and its Remediation (ECR21)
Chandigarh, India, October 18–22, 2021

Comparative Study on Fly Ash based AAM Concrete with GGBS, Rice Husk Ash and Sugarcane Bagasse Ash

V. Poornima[1], K. Vasanth Kumar[2] and P.P. Hridhi Nandu[3]

ABSTRACT

One of the main staples in the construction industry is Portland cement-based concrete, which currently contributes to 8% of global CO_2 emissions during cement production. In this respect, the geopolymer technology shows considerable promise for application in the concrete industry as an alternative binder to Portland cement. Still, there has not been a significant shift away from the use of Portland Cement due to the demand for additional necessary precautions and the requirement of high temperature to bring about rapid strength gain. The mechanical and durability properties of class F fly-ash (FA), based Alkali Activated Concrete (AAM) with varying percentages of Ground Granulated Blast Furnace Slag (GGBS), Rice Husk Ash (RHA), and Sugarcane Bagasse Ash (SCBA) cured at ambient weather conditions are discussed. The results of the tests further indicate the possibility of using ambient cured AAM Concretes for construction purposes.

Keywords: Fly Ash, Ground Granulated Blast Furnace Slag (GGBS), Geopolymer Concrete (GPC), Rice Husk Ash (RHA), Sugarcane Bagasse Ash (SCBA), Agricultural waste, Sustainable Concrete, Concrete, Alternative Construction Materials.

[1] *Corresponding Author.*
Assistant Professor, Department of Civil Engineering, Amrita School of Engineering, Coimbatore, India.
E-mail: v_poornima@cb.amrita.edu
[2] *Presenting & corresponding author.*
Undergraduate Student, Department of Civil Engineering, Amrita School of Engineering, Coimbatore, India.
E-mail: k.vasanth@protonmail.com
[3] *Presenting & corresponding author.*
Undergraduate Student, Department of Civil Engineering, Amrita School of Engineering, Coimbatore, India.
E-mail: nhridhi001@gmail.com

F-EIR Conference 2021 on
Environment Concerns and its Remediation (ECR21)
Chandigarh, India, October 18–22, 2021

Trends and Advances in Pesticide's Remedial Approaches Applied on Agriculture Produces for Food Safety at Consumer Level: A Review

Chetan Chauhan[1] and Shanta Kumari[2]

ABSTRACT

Pesticide is one the inevitable input in the modern agriculture since green revolution era but its non judicious and overuse severely affecting the nutrient profile and quality of agricultural produces due to presence of pesticides residues above MRLs in food chain leading to alarming health and environment concern. And situation further aggravated when agriculture produces consumed the before the waiting period. Therefore, always we have risk coupled during the consumption of agricultural produces due to presence of pesticide residues so consumers can take certain prompt action to lower their exposure to pesticide residues through its removable from agricultural produces. Although, simple household procedures to advance methodologies have been explored for the mitigation and removable of pesticides and their residues from agriculture produces. Currently there are effective approaches reported in literature through the application of various modern technologies such as ozonation, ultrasonication, electrolyzed water (EW), cold plasma (NTP), high hydrostatic pressure (HHP), radiolytic method etc for removable of pesticides from agricultural produces However, lesser number of advance methods are available and accessible to common public due their high installation, maintenance and running cost. The present article reviewed the various conventional to advance risk mitigation approaches for pesticides and their residues from vegetables and fruits for safe food consumption (without any physical and chemical side effect on the agricultural produces) in order to address pesticide associated concerns.

Keywords: EW, HHP, NTT, NTP, MRL.

---

[1] *Presenting & corresponding author.*
Division Chemistry, University Institute of Sciences, Chandigarh University, India.
E-mail: chauhan.chetan123@gmail.com
[2] Department of Economics, Eternal University, India.

F-EIR Conference 2021 on
Environment Concerns and its Remediation (ECR21)
Chandigarh, India, October 18–22, 2021

Catalytic Activity of Nano Mesoporous Lanthanum Oxide for Liquid Phase Benzylation of O-Xylene

Marymol Moothedan[1] and K.B. Sherly[2]

ABSTRACT

Nano mesoporous lanthanum oxide (La_2O_3) was synthesized by starch template method. The physicochemical characterization of La_2O_3 was performed by versatile techniques like TGA-DTG/DSC, XRD, FT-IR, TEM and Surface area and porosity measurements. Nano mesoporous La_2O_3 prove to be an efficient catalyst for Friedel-Crafts benzylation of o-xylene (o-Xyl) with benzyl chloride (BC) in a batch reactor at atmospheric pressure. The effect of reaction temperature, catalyst loading and reaction time on benzylation of o-Xyl was studied at different o-Xyl/BC molar ratio. High BC conversion with good 3, 4-dimethyl diphenyl methane (3, 4-DMDPM) selectivity was achieved with 0.1 g catalyst at 433 K for 2 h of reaction time and 1:1 reactant molar ratio. 3, 4-DMDPM was obtained as the major product regardless of temperature, time and reactant molar ratio. The BC conversion and selectivity found to increases with reaction temperature, catalyst weight and reaction time and decrease with increase with reactant molar ratio.

Keywords: Lanthanum Oxide, Liquid Phase Benzylation, 3, 4-DMDPM, Mesoporous, Starch Template.

[1] *Presenting author.*
Regional Research Centre of M.G. University, Mar Athanasius College, Kothamangalam, Kerala 686666, India.
E-mail: marymolmoothedan@gmail.com
[2] Regional Research Centre of M.G. University, Mar Athanasius College, Kothamangalam, Kerala 686666, India, India. *E-mail:* sherlykb@gmail.com

F-EIR Conference 2021 on
Environment Concerns and its Remediation (ECR21)
Chandigarh, India, October 18–22, 2021

Synthesis, Characterization and Phosphate Adsorption Studies of Nano Mesoporous Lanthanum Oxide

Marymol Moothedan[1] and K.B. Sherly[2]

ABSTRACT

In this study nano mesoporous lanthanum oxide (La_2O_3) was synthesized by chitosan template method. Formation, crystalline character, stability, particle size and porosity of La_2O_3 were investigated from XRD, FT-IR, TG/DTG and BET analysis. Synthesized La_2O_3 showed good phosphate adsorption capacity of 266.7 mg g^{-1}. The uptake of phosphate increased with increasing initial phosphate concentration and reaction time until equilibrium was achieved. Maximum adsorption capacity was obtained for 480 ppm initial phosphate solution with 1 h of reaction time. The adsorption isotherm data correlated well with Redlich-Peterson isotherm model and kinetic data with Pseudo second-order kinetic model. Nano mesoporous La_2O_3 after phosphate adsorption was characterized and probable mechanisms for the phosphate adsorption was also proposed.

Keywords: Chitosan, Lanthanum oxide, Mesoporous, Phosphate.

Graphical Abstract

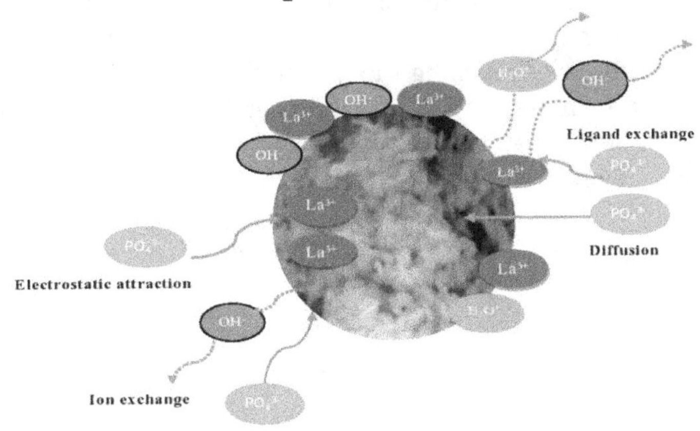

---

[1] *Presenting & corresponding author.*
Regional Research Centre of M.G. University, Mar Athanasius College, India.
E-mail: marymolmoothedan@gmail.com
[2] *Corresponding author.*
Regional Research Centre of M.G. University, Mar Athanasius College, India. *E-mail:* sherlykb@gmail.com

F-EIR Conference 2021 on
Environment Concerns and its Remediation (ECR21)
Chandigarh, India, October 18–22, 2021

Durability of Treated Recycled Aggregate based Concrete

Sathvik Reddy Nimma[1] and Mukund Lahoti[2]

ABSTRACT

Building materials constitute about half of all the materials used by us and about half the solid waste generated worldwide. Recycled construction and demolishment aggregates (CDA) can be an invaluable source of construction material. In India, if CDA waste is quantified, it will be more than all the other types of solid wastes put together. Even though the gravity of the problem is understood and usage of recycled aggregates in new construction is being encouraged throughout the world, the structural use of recycled aggregate concrete is limited because of its inferior mechanical strength, and durability performance. Hence, there is a need to treat these aggregates and produce better quality aggregates suitable for good structural grade concrete. The present work focuses on the study and comparison of the effects due to different treatment techniques of recycled aggregates. Effective treatment techniques separate the composite of natural aggregate and the mortar sticking to it. Different RA treatment methods such as – treatment of recycled fine aggregates with tannic acid, soaking the RA in acetic acid with the mechanical scraping of the mortar, accelerated carbonation treatment, soaking in acetic acid with accelerated carbonation and soaking in lime with accelerated carbonation, crushing the aggregates to increase the self-cementing capacity (for improving the bearing capacity as the recycled aggregates are more prone to secondary cement setting), etc. are considered for the study. Durability performance - absorption (of water), acid attack, permeability, chloride attack, and carbonation for treated aggregates-based concrete are discussed.

Keywords: Recycled Aggregates, Recycled Aggregate Concrete, Treatment.

---

[1] *Presenting author.*

Undergraduate Student, Department of Civil Engineering, Birla Institute of Technology and Science, Pilani, India. *Email:* f20170993@pilani.bits-pilani.ac.in

[2] *Corresponding author.*

Assistant Professor, Department of Civil Engineering, Birla Institute of Technology and Science, Pilani, India. *E-mail:* mukund.lahoti@pilani.bits-pilani.ac.in

F-EIR Conference 2021 on
Environment Concerns and its Remediation (ECR21)
Chandigarh, India, October 18–22, 2021

A Comparative Study on the LPG Sensing Performance of Spray Pyrolyzed SnO_2 Thin Films Prepared from Air Blast and Ultrasonic Atomizers

Andrew Simon George[1], Deepa S.[2], Radhika Keshavan[3] and Keerthana Bose[3]

ABSTRACT

In the context of rapid industrialization and globalization, detection of pollutant, toxic, refining, combustive and process gases is important for system and process control, safety monitoring and environmental protection. Spray pyrolysed semiconductor oxide thin films such as SnO_2 have an excellent potential for gas sensor applications due to their high capacity to adsorb gaseous molecules and promote their reactions (usually their oxidation) while showing changes in their surface conductivity. It is a well established fact that controlled crystallite size along with a favourable morphology enhances sensor response with improved response and recovery time. In this context, atomization of liquids and the atomization scheme employed during spray pyrolysis plays a crucial role. The present work is an attempt to study the effect of atomizer in shaping the microstructural, morphological as well as optical properties of SnO_2 thin films. Two different atomizers are chosen for spray pyrolysis - one air-blast and the other ultrasonic type. Keeping the same deposition conditions, SnO_2 thin films are prepared using the two atomizers. The microstructural investigation is carried out using X-Ray Diffraction (XRD) supported by Transmission Electron Microscopy (TEM). The UV-Vis absorbance spectra shows a significant reduction in bandgap for the samples prepared via ultrasonic atomizer in comparison to those from air blast atomizer. The variation in surface features are analyzed by FESEM images. The LPG sensing property of the samples is analyzed using Keithley 2400 source meter. The enhanced sensing performance of the sample prepared by air blast atomizer is substantiated by the modified microstructural and morphological features.

Keywords: SnO_2, *Thin Films, LPG Sensing, Air blast Atomizer, Ultrasonic Atomizer.*

---

[1] *Corresponding author.*
Research Centre in Physics, Mar Athanasius College (Autonomous), M.G. University, Kothamangalam, India.
[2] *Presenting & corresponding author.*
Research Centre in Physics, Mar Athanasius College (Autonomous), M.G. University, Kothamangalam, India.
E-mail: deepa.venukumar@gmail.com
[3] Research Centre in Physics, Mar Athanasius College (Autonomous), M.G. University, Kothamangalam, India.

CO$_2$ Sensing Properties of Spray Pyrolysed Ce Doped SnO$_2$ Thin Film

A.R. Hariprasad[1], Andrew Simon George[2] and S. Deepa[3]

ABSTRACT

Metal oxide semiconductors used as gas sensors are based on the dependence of their electrical properties in the ambient gas concentration. The structural properties of thin films have significant effect on the electrical and gas sensing properties, as the shift from bulk to the thin film state may cause radical change in its material behaviour. In this context, spray pyrolysed Tin Oxide is one of the most promising candidate. Defect composition and the contribution of different kinds of oxygen vacancies (OVs) have a crucial role on gas sensing properties of SnO$_2$. It has been reported that doping with rare-earth elements produces significant changes in structural, morphological and optical properties and thereby improving their gas sensing characteristics. In this work, nano particulate SnO$_2$ thin films are prepared using Cerium (Ce) as a dopant. Structural characterization is done via X-Ray Diffraction (XRD), which confirms the formation of consistent films with tetragonal cassiterite structure. The crystallite size is found to lie in the nano regime of 23–35 nm with optimal doping (1 wt.%) showing improved crystallinity. Surface features are analyzed using Field Emission Scanning Electron Micrograph (FESEM). UV-Vis absorbance spectra confirm enhancement in optical properties upon Ce doping. CO$_2$ sensing properties of the prepared films are investigated using Keithley 2400 source meter. 1 wt.% of Ce doped SnO$_2$ thin film shows maximum sensor response of 90.8% with improved response time.

Keywords: SnO$_2$, Thin Films, Ce doped, CO$_2$ Gas Sensing.

[1] Department of Physics, Indira Gandhi College of Arts & Science, India.

[2] *Corresponding author.*

Research Centre in Physics, Mar Athanasius College (Autonomous), Kothamangalam College P.O, Kothamangalam, Kerala 686 666, India. *E-mail:* andrewsimongeorge@gmail.com

[3] *Presenting & corresponding author.*

Research Centre in Physics, Mar Athanasius College (Autonomous), Kothamangalam College P.O, Kothamangalam, Kerala 686 666, India. *E-mail:* deepa.venukumar@gmail.com

F-EIR Conference 2021 on
Environment Concerns and its Remediation (ECR21)
Chandigarh, India, October 18–22, 2021

Comparative Life Cycle Assessment of Innovative Hazardous Waste Valorizing Bricks with Conventional Bricks

Gaurav Tyagi[1], Srikanta Routroy[2], Anupam Singhal[3], Dipendu Bhunia[4] and Mukund Lahoti[5]

ABSTRACT

Production of stainless steel generates hazardous Nickel-Chromium Plating Sludge (NCPS) containing high concentrations of heavy metals. Unavoidably, manufacturers prefer the expensive method of cement-based landfilling over commercially and technologically impractical methods of recovering heavy metals from sludge. This paper evaluates the environmental impacts of valorizing NCPS in the novel NCPS-fly ash-clay fired bricks towards a sustainable solution. Life cycle assessment (LCA) is conducted for conventional clay bricks and the test bricks. Test bricks use only 50% clay unlike conventional clay bricks. The rest is constituted by sandy soil, fly ash, and NCPS- whose optimum proportions are reached experimentally. While sandy soil's addition lowers the fertile top soil's requirement for clay extraction, fly ash also acts as an internal fuel due to unburnt carbon, reducing coal consumption substantially. LCA analysis is performed utilizing ReCiPe midpoint and endpoint methods in open LCA software using primary data obtained during laboratory experiments and secondary data from the Ecoinvent version 3.1 database. A sensitivity analysis is carried out to analyze the effect of transportation of raw materials. LCA results indicate that coal combustion and topsoil extraction are prominent sources of harmful environmental impacts in the manufacture of conventional bricks. The 'cradle to gate' methodology is employed for the analysis. The superiority of the novel bricks over conventional bricks is demonstrated by quantifying emissions and energy use and assessing the environmental impacts such as agricultural land occupation, climate change, fossil depletion, particulate matter formation, and human toxicity.

Keywords: Nickel-Chromium Plating Sludge, Life Cycle Assessment, Bricks.

[1] *Presenting author.* Research Scholar, Department of Civil Engineering, Birla Institute of Technology and Science, India. *E-mail:* p20180419@pilani.bits-pilani.ac.in

[2] *Corresponding author.* Professor, Department of Mechanical Engineering, Birla Institute of Technology and Science, India. *E-mail:* srikantaroutroy@gmail.com

[3] Professor, Department of Civil Engineering, Birla Institute of Technology and Science, India. *E-mail:* anupam.singhal@pilani.bits-pilani.ac.in

[4] Associate Professor, Department of Civil Engineering, Birla Institute of Technology and Science, India. *E-mail:* dbhunia@pilani.bits-pilani.ac.in

[5] Assistant Professor, Department of Civil Engineering, Birla Institute of Technology and Science, India. *E-mail:* mukund.lahoti@pilani.bits-pilani.ac.in

F-EIR Conference 2021 on
Environment Concerns and its Remediation (ECR21)
Chandigarh, India, October 18–22, 2021

Measurement of Aggregates Breakage of Asphalt Mixtures with Construction and Demolition Waste

Abbaas I. Kareem[1], Qais Sahib Banyhussan[2] and Hayder Mohammed Al-Taweel[3]

ABSTRACT

In this paper, the effect of mechanical mixing and compaction on the physical properties of fragile CDW used in asphalt mixtures evaluated. The paper investigates the change in aggregates gradation of CDW-asphalt mixtures after mechanical mixing and compaction. Asphalt specimens containing 0% CDW, 25% CDW, 50% CDW, and 75% CDW were prepared according to Marshall procedure. The particle size distribution test was carried out to measure the broken rate of CDW due to mechanical mixing and compacting. Three samples of each mix were tested and the average value was considered. The results reveal that the broken CDW were able to pass sieve 4.75 mm. The change in aggregates gradation was mainly due to the breakage of CDW during mixing and compaction. The breakage of CDW led to a decrease in the dosage of coarse CDW presented in aggregates retained on 4.75 mm sieve and an increase in the dosage of fine CDW presented in aggregates that passed the 4.75 mm sieve. The results also indicate that the broken rate increased as the percentage of CDW increased in the mix. The breakage of CDW has promoted a change in the density of combined mineral aggregates, the density of combined coarse aggregates, and the density of combined fine aggregates of CDW-mixtures. The breakage also contributes to a change in water absorption of combined coarse aggregates, and that measured for combined fine aggregates of CDW-asphalt mixtures. The results indicate the need for measuring the

[1] *Presenting & corresponding author.*
Curtin University, Department of Civil Engineering, Perth, Australia.
Mustansiriyah University, Highway and Transportation Engineering Department, Bab Almoatham, Baghdad, Iraq.
E-mail: abbaaskareem@uomustansiriyah.edu.iq; a.kareem@postgrad.curtin.edu.au
[2] Mustansiriyah University, Highway and Transportation Engineering Department, Bab Almoatham, Baghdad, Iraq.
E-mail: qaisalmusawi@uomustansiriyah.edu.iq
[3] Mustansiriyah University, Highway and Transportation Engineering Department, Bab Almoatham, Baghdad, Iraq.
E-mail: hayder.moh.altaweel@uomustansiriyah.edu.iq

amount of broken CDW passing 4.75 mm sieve and evaluate its effect on physical properties of aggregates of CDW-asphalt mixtures.

Keywords: *Aggregates Breakage, Asphalt Mixtures, Construction and Demolition Waste, Mechanical Compaction, Broken Rate, Fragile Aggregates.*

F-EIR Conference 2021 on
Environment Concerns and its Remediation (ECR21)
Chandigarh, India, October 18–22, 2021

Investigation of Equilibrium CO_2 Solubility in 35 wt % Aqueous 1-(2-aminoethyl) Piperazine (AEP) and Performance Study over Monoethanolamine for CO_2 Absorption

Anirban Dey[1], Sweta C. Balchandani[2], Bharti Saini[2] and Sukanta Kumar Dash[2]

ABSTRACT

Due to the growing concern of global warming in the environment, various researches are going on across the world in the development of more efficient technologies in order to combat this issue. Absorption by alkanolamine solvents is the most promising technology widely used for the capture of CO_2 from industrial flue gas stream. In the present study, experimental CO_2 solubility in aqueous solution of 1-2-(aminoethyl) piperazine (AEP) was investigated at concentration of 35 wt % .The solubility was measured in a high pressure solubility cell at temperatures 308.15,318.15 and 328.15 K, over a CO_2 pressure range of (2–200 kPa). The effect of equilibrium CO_2 pressure, temperature and AEP concentration on CO_2 loading were examined. The equilibrium solubility data were further correlated by semi-empirical model. In addition to this a parametric study was also conducted in Aspen Plus platform to investigate the performance of AEP over conventional solvent MEA in terms of % CO_2 removal.

Keywords: Solubility, Carbon-dioxide, AEP, Semi-empirical Model.

---

[1] *Presenting & corresponding author.*
CO_2 Research Group, Department of Chemical Engineering, Pandit Deendayal Energy University, Gandhinagar 382007, India. *E-mail:* anirban.dey@sot.pdpu.ac.in
[2] CO_2 Research Group, Department of Chemical Engineering, Pandit Deendayal Energy University, Gandhinagar 382007, India.

F-EIR Conference 2021 on
Environment Concerns and its Remediation (ECR21)
Chandigarh, India, October 18–22, 2021

Hydrodynamic Cavitation: Its Optimization and Potential Application in Treatment of Pigment Industry Wastewater

Sanyukta Singh[1] and Shrikant Randhavane[2]

ABSTRACT

Paper presents the application of Hydrodynamic Cavitation technique in reduction COD and its various fractions from the Pigment Industry wastewater. Orifice plate with dimension of single hole of 1.5 mm is used in this study. The sample were analyzed at different pressure of 2 bar, 3 bar, 4 bar, 5 bar and 6 bar at different time intervals of 30 min, 60 min, 90 min, 120 min and 180 min. The optimum time and inlet pressure for maximum COD reduction is obtained at 5 bar and 120 minutes. The samples are analyzed in two stages. In stage first stage, samples are treated with hydrodynamic cavitation only and filtered thereafter. Both the unfiltered and filtered samples were analyzed for various parameters such as COD (filtered and unfiltered), BOD (filtered and unfiltered), TDS and pH. In stage two, the samples are treated with aeration for optimum time followed by hydrodynamic cavitation. The optimum time of aeration was obtained at 36 hrs. Response Surface Methodology was used to check the sensitivity of the hydrodynamic cavitation set up. Optimization using RSM provided optimum results with desirability 0.798 when pressure was at 5 bar and time of reaction was 120 minutes.

Keywords: *Hydrodynamic Cavitation, Aeration, Refractory COD, Soluble COD, Pigment Wastewater.*

[1] *Presenting author.*
Environmental Engineer, Engineers India Limited, India. *E-mail:* sanyuktakshatriya@gmail.com
[2] *Corresponding author.*
Assistant Professor, Department of Civil Engineering, SVKM's Institute of Technology, Dhule 424001, India.
E-mail: srandhavane@gmail.com

F-EIR Conference 2021 on
Environment Concerns and its Remediation (ECR21)
Chandigarh, India, October 18–22, 2021

Effectiveness of Unburnt Rice Husk, Fly Ash, and Zeolite as a Substitute of Cement in Concrete

Mahadeb Das[1], Suman Kumar Adhikary[2] and Zymantas Rudzionis[3]

ABSTRACT

This experimental study analyzed the impacts of unburnt rice husk, fly ash, and zeolite on the strength of glass powder added concrete. At the first stage of the study up to 30% mass of fine aggregates was replaced by glass powder to understand the impact of the glass powder on the strength of concrete. Study results show that increasing the concentration of glass powder tends to decrease the compressive strength of the concrete. The relative decrease in compressive strength with 10% and 30% content of glass powder was measured at 9.5% and 13.3%, respectively. Furthermore, at a 10% constant concentration of glass powder several concrete specimens were prepared using unburnt rice husk, fly ash, and zeolite powder as replacement of cement (up to 15%). Study results indicate that the use of unburnt rice husk, fly ash, and zeolite with the combination of glass powder decreases the compressive strength of concrete. The lowest reduction rate was observed for fly ash added concrete, while the addition of unburnt rice husk shows the highest reduction in compressive strength.

Keywords: Glass Powder, Fly Ash, Zeolite, Unburnt Rice Husk, Compressive Strength.

[1] Siliguri Institute of Technology, Sukna 734009, India. *E-mail:* mahadebster@gmail.com

[2] *Presenting & corresponding author.*
Faculty of Civil Engineering and Architecture, Kaunas University of Technology, Kaunas, LT - 44249, Lithuania. *E-mail:* sumankradk9s@gmail.com

[3] Faculty of Civil Engineering and Architecture, Kaunas University of Technology, Kaunas, LT - 44249, Lithuania. *E-mail:* zymantas.rudzionis@ktu.lt

F-EIR Conference 2021 on
Environment Concerns and its Remediation (ECR21)
Chandigarh, India, October 18–22, 2021

Drought Identification and Analysis of Precipitation Trends in Beed District, Maharashtra

Satish G. Taji[1] and Abhijeet P. Keskar[2]

ABSTRACT

Drought is a complex natural hazard with slow-onset and can occur over months or even years. Typically, a drought index is used to quantify the effect of drought which is based on various meteorological parameters, most importantly precipitation. However, the changing pattern of the rainfall is now a concerning issue and therefore, trend analysis of the precipitation data has become a vital part of the study, as it gives important information of the possibility of changes before prediction and analysis of natural hazards like drought. In the present study, spatial and temporal assessment of meteorological drought has been carried out for the Beed district of the Marathwada region which is known as a drought-prone area. Initially, trend analysis of 34 years of precipitation data has been carried out using non-parametric tests Man-Kendall and Sen's slope estimator. Standardized precipitation index (SPI) values of different timescales (1, 3, 6, 9, and 12-monthly) were used to identify and analyze the drought events. For understanding the spatial variation of SPI12 over the region, SPI maps have been prepared using GIS. The study has identified total six drought events (1985–86, 1987–88, 1992–93, 2001–02, 2005–06, and 2011–12) from 1979 to 2013. The results of the study show that Jamkhed, Ashti, Majalgaon, and Ambejogai are the most affected areas, whereas, Georai and Beed are the least affected.

Keywords: Drought, SPI, Trend Analysis, GIS, Non-parametric Test.

[1] *Presenting & corresponding author.*
Assistant Professor, Civil Engineering Department, SVKM's Institute of Technology, Dhule, India.
E-mail: satishtaji777@gmail.com
[2] Assistant Professor, Civil Engineering Department, Terna Engineering College, India.
E-mail: keskarabhijeet@gmail.com

F-EIR Conference 2021 on
Environment Concerns and its Remediation (ECR21)
Chandigarh, India, October 18–22, 2021

Restoring Urban Green Cover of Chennai City: An Ecological Approach

A. Meenatchi Sundaram[1]

ABSTRACT

Green cover has bigger role in linking Atmosphere, lithosphere, hydrosphere, biosphere, and anthroposphere on the earth. Reduction of green space is the main cause for most of environmental issues we are facing. Impervious land covers, – the one such alteration, as well characteristics features of the urban area becomes the prominent environmental indicators for wide range of urban environmental issues (Chester L.Arnold *et al.*, 1996, Stephan Pauleit *et al.*, 2000). From the literatures it has been envisaged that implementing appropriate green cover plan could mitigate most of the current urban environmental problems, particularly air pollution, water scarcity and urban heat island; the three major issues that trembles the urban quality of life. It is also found that adapting structure and function based spatial strategies of landscape ecology in the urban green cover plan could solve many ecological issues in the cities. Mainly protection of ecologically sensitive areas; connecting the fragmented natural areas; and ease the flow or movement of energy, matter, and species across the urban landscape. Through adapting the patches, corridor and matrix based spatial concepts this study developed four level strategies for the urban green cover restoration of Chennai city.

Keywords: *Ecosystem Service, Urban Green Space, Natural Process, Pollution Removal, Surface Runoff, Spatial Index.*

---

[1] *Presenting & corresponding author.*
Associate Professor, Department of Architecture, NIT, Trichy, India. *E-mail:* meenatchi@nitt.edu

F-EIR Conference 2021 on
Environment Concerns and its Remediation (ECR21)
Chandigarh, India, October 18–22, 2021

Effect of GGBS on Fly Ash based Geopolymer Mortar at Ambient and Heat Curing

Abhishek Sharma[1], Paramveer Singh[2] and Kanish Kapoor[3]

ABSTRACT

This study investigated and showed the effect of incorporation of Ground Granulated Blast furnace Slag (GGBS) content in Fly Ash (FA)-based geopolymer mortar by assessing the strength properties of geopolymer mortar in both ambient (25°C) and heat (60°C) curing. The percentage of GGBS was varied from 0% to 50% with FA in geopolymer mortar. Sodium silicate (SS) and sodium hydroxide (SH) solutions were used to activate the main binders. For all the mixtures the alkali/binder ratio, molarity (M), and Sodium silicate (SS)/sodium hydroxide (SH) ratio were fixed to 0.45,12 M, and 2, respectively. The fresh property in terms of slump flow and other properties like compressive strength, water absorption, and dry density of the geopolymer mortar were determined. The result of this study showed that the inclusion of GGBS content in FA-based geopolymer mortar significantly increased the compressive in both the curing conditions. Maximum compressive strength of 44.33 MPa was obtained when 50% GGBS was added to the mixtures at a curing temperature of 60°C.

Keywords: Geopolymer Mortar, Compressive Strength, Water Absorption, Ambient Curing.

[1] Department of Civil Engineering, Dr. B.R. Ambedkar National Institute of Technology, Jalandhar, India.
E-mail: abhisharma1216@gmail.com
[2] *Presenting & corresponding author.*
Department of Civil Engineering, Dr. B.R. Ambedkar National Institute of Technology, Jalandhar, India.
E-mail: paramveers.ce.19@nitj.ac.in
[3] Department of Civil Engineering, Dr. B.R. Ambedkar National Institute of Technology, Jalandhar, India.
E-mail: kapoork@nitj.ac.in

F-EIR Conference 2021 on
Environment Concerns and its Remediation (ECR21)
Chandigarh, India, October 18–22, 2021

Ultrasound-Assisted Oxidative Desulfurization of Base Oil with a Dimethyl Imidazolium Dimethyl Phosphate as a Catalyst/Extractant

Komal G. Desai[1] and Swapnil A. Dharaskar[2]

ABSTRACT

Owing to the increasing use of oils and global air production, a comprehensive development towards the production of ultra-clean sulfur-free oil is increasing day by day. Production of ultra-pure oils can be obtained by some limited and the most integrated techniques. Ultrasound-assisted oxidative desulfurization (UAODS) process possessing appropriate characteristics with regards to the oxidant-ionic liquid (IL) system. IL, 1, 3-dimethyl imidazolium dimethyl phosphate ([MMIM]DMP) was synthesized and characterized by FT-IR, NMR, TGA, and DSC analysis. Using n-octane as a synthetic oil (SO) with dibenzothiophene (DBT) as a sulfur source under various sonication conditions is the attempt of this work. The influential parameters such as sonication time and temperature, mass ratio of IL to SO (m(IL/SO)), and mole ratio of oxidant to sulfur (n(O/S)) are observed and optimized using Response Surface Methodology-Box Behnken Design (RSM-BBD). The maximum desulfurization efficiency of 99.04% is obtained under the sonication time of 15 minutes with n(O/S) of 6 in an acidic pH atmosphere. In this research, m(IL/SO) of 1/1 represented the most influential role in the oxidation reaction. Oxidation reaction followed the pseudo-1st-order reaction kinetics. The IL was recycled conveniently and reused for 8 cycles without much decrease in the activity. The mechanism behind the extraordinary performance of the ionic liquid under sonication was briefly studied using the DFT study. The maximum interaction energy between the ionic liquid and DBT and DBTO$_2$ was owing to the CH-Π, Π-Π, and H-bonding interactions. The efficacy of [MMIM]DMP and the impact of ultrasound irradiation on the desulfurization of real base oil with an initial sulfur content of 1123 ppm were checked using only extraction, extractive oxidation with and without ultrasound. The results obtained proved that the UAODS process using [MMIM]DMP in the presence

[1] *Presenting author.*
Department of Chemical Engineering, Pandit Deendayal Energy University, Gandhinagar 382007, India.
E-mail: desaikomal91@gmail.com; komal.dphd18@sot.pdpu.ac.in
[2] *Corresponding author.*
Department of Chemical Engineering, Pandit Deendayal Energy University, Gandhinagar 382007, India.
E-mail: swapnildharaskar11@gmail.com; Swapnil.dharaskar11@sot.pdpu.ac.in

of an equimolar mixture of hydrogen peroxide and acetic acid can achieve the maximum desulfurization performance and reduced the sulfur content of the base oil <10 ppm. This implies that the efficacy of the UAODS process using [MMIM]DMP for the industrial application.

Keywords: *1, 3-dimethyl Imidazolium Dimethyl Phosphate, Phase Transfer Catalyst, Ionic Liquid, Ultrasound-Assisted Oxidative Desulfurization, RSM-BBD, DFT Study.*

Graphical Abstract

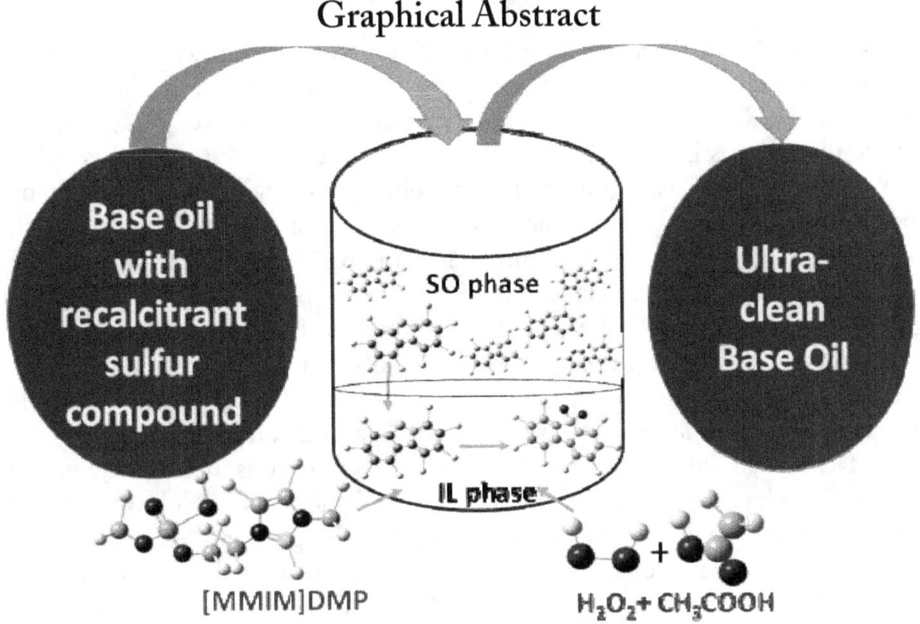

F-EIR Conference 2021 on
Environment Concerns and its Remediation (ECR21)
Chandigarh, India, October 18–22, 2021

The Impact of the COVID-19 Pandemic on Shopping Trips Behavior in Urban Areas

Abeer Khudhur Jameel[1], Wafaa Kh. Luaibi[2] and Iman Alghalibi[3]

ABSTRACT

The impact of COVID-19 pandemic is highly considered in the recent time because of its significant role in changing the behavior of trips. However, its effect on the shopping trips according to the shopping types is rarely considered. Therefore, this paper aims to investigate the changes in the behavior of shopping trips and their control factors during the COVID-19 pandemic according to the shopping types. The considered shopping types are food and drinks, cosmetics, clothes and shoes, electronic devices, dolls and children products, electricity devices and furniture. The time of COVID-19 pandemic is divided into six stages stating from December 2019 to the recent time. The collected data are categorized into two groups. The first group represents the changes in the behavior of shopping trips including frequency, mode choice, distance of trip, and time of trip. The second group represents the socio demographic factors such as age, gender, marital states, income level, education level, and car ownership. It also includes the experience in using electronic facilities and location factor. The questionnaire form is used as a method of data. The results show increase in the rate of food, electronic devices and dolls shopping despite the remarkable decline in the rate of other shopping types between March 2020 and October 2020, after that the behavior has been similar to the behavior of shopping trips before COVID-19 pandemic. However, the frequency of using e-shopping and using walking in short shopping trips is still higher than before pandemic even in post-pandemic period. This encourages using avoid-shift strategy to promote sustainable transportation system and cities.

Keywords: *COVID-19, Shopping Trips, Trip Behavior, Control Factors, E-shopping, Sustainable Transport Mode.*

[1] *Presenting & corresponding author.*
Assistant Professor, Highway and Transportation Engineering Department, College of Engineering, Mustansiriyah University, Baghdad, Iraq. *E-mail:* abeer_khudhur@uomustansiriyah.edu.iq
[2] Lecturer, Highway and Transportation Engineering Department, College of Engineering, Mustansiriyah University, Baghdad, Iraq. *E-mail:* wleabi@uomustansiriyah.edu.iq
[3] Assistant Lecturer, Highway and Transportation Engineering Department, College of Engineering, Mustansiriyah University, Baghdad, Iraq. *E-mail:* imanalghalibi@uomustansiriyah.edu.iq

F-EIR Conference 2021 on
Environment Concerns and its Remediation (ECR21)
Chandigarh, India, October 18–22, 2021

Health Monitoring of Bacterial Concrete Structure under Dynamic Loading Using Electro Mechanical Impedance Technique: A Numerical Approach

Krishna Kumar Maurya[1], Anupam Rawat[2] and Rama Shanker[3]

ABSTRACT

Health monitoring of structures using techniques based on the smart material is an innovative concept that is exploding technological revolutions in the field of civil engineering. The Electro-mechanical impedance (EMI) technique is used for structural health monitoring (SHM) and to investigate the damages in the structures. The bacterial incorporation in concrete produces calcite material through metabolism process in presence of moisture and carbon di-oxide, which improves the mechanical properties of concrete. Hence, its application in construction of the buildings will improve the health of structures. In this research paper, dynamic behaviour of the bacterial concrete was investigated numerically. The beams of size $700 \times 150 \times 150$ mm of bacterial concrete and standard concrete were modelled using finite element based package ANSYS19. The beam of bacterial concrete was simulated as per the characteristics of the materials produced after the bacterial metabolism reactions. The EMI technique was applied to investigate the health of these beams. Admittance (conductance and susceptance) signatures were determined using piezo-ceramic lead zirconate titanate (PZT) sensors installed at mid-point on the top surface of concrete beams. The beams were exposed to dynamic loading and the intensity of dynamic loading was increased in four sub-steps. For the quantification of strength development in the concrete beam, the root mean square deviation (RMSD) statistical index had been applied. It was observed that the bacterial concrete beam has more resistance to the dynamic loading.

Keywords: *Structural Health Monitoring, Bacterial Concrete Beam, EMI Technique, Sensor System, Dynamic Loadings, Statistical Index.*

---

[1] *Presenting & corresponding author.*
Ph.D. Research Scholar, Department of Civil Engineering, Motilal Nehru National Institute of Technology Allahabad, Prayagraj 211004, India. *E-mail:* krish01990@gmail.com
[2] Assistant Professor, Department of Civil Engineering, Motilal Nehru National Institute of Technology Allahabad, Prayagraj 211004, India. *E-mail:* anupam@mnnit.ac.in
[3] Associate Professor, Department of Civil Engineering, Motilal Nehru National Institute of Technology Allahabad, Prayagraj 211004, India. *E-mail:* ramashanker@mnnit.ac.in

F-EIR Conference 2021 on
Environment Concerns and its Remediation (ECR21)
Chandigarh, India, October 18–22, 2021

Assessing the Sustainable Valorisation and Decontamination of Wine Wastewater via Anaerobic Tests with Subsequent Dynamic Adsorption Studies

Arunima Nayak[1], Brij Bhushan[2], Marco Ortiz-Cabrera[3] and Xavier Flotats[3]

ABSTRACT

Anaerobic digestion (AD) via biochemical methane (BMP) tests and sequencing batch reactor tests (ASBR) with subsequent adsorption tests was conducted on a laboratory scale to assess the decontamination and bio-methanization efficiency of a real wine wastewater (WWW).

BMP tests conducted on WWW at 30°C for 58 days using inoculum having previous adaptability to the winery polyphenols (TPP) resulted in higher biodegradability (93%), higher polyphenols degradation (86%) and a near complete removal of chemical oxygen demand (COD). ASBR trials on the same WWW revealed a bio-methane yield of 63%. Also the final effluent (PWE) showed a COD reduction of 93%; with a near elimination of volatile fatty acid (VFA); but the final TPP was recorded as 150 mg/L. Adsorption studies carried onto a fixed bed porous adsorbent developed from exhausted grape pomace (GPAC) enabled a 67% removal of TPP from PWE under optimum bed height and flow conditions. Studies reveal that ASBR tests conducted with inoculum having previous adaptability to winery polyphenols followed by adsorption onto fixed bed of grape pomace activated carbon could result in significant decontamination and high organic loading in WWW could be efficiently valorised to significant biogas yields.

Keywords: Adsorption, Anaerobic Sequence Batch Reactor, Biochemical Methane Potential, Biogas, Polyphenols Removal, Wine Wastewater.

---

[1] *Corresponding author.*

Graphic Era University, Dehradun 248002, India. *E-mail:* arunima_nayak@yahoo,com

[2] *Presenting author.*

Graphic Era University, Dehradun 248002, India.

[3] Universitat Politècnica de Catalunya, UPC-BarcelonaTECH, Parc Mediterrani de la Tecnologia, Building D4, E-08860 Castelldefels, Barcelona, Spain.

F-EIR Conference 2021 on
Environment Concerns and its Remediation (ECR21)
Chandigarh, India, October 18–22, 2021

Low Energy, Long Sustainable and High-Speed FIR Filter based on Truncated Multiplier with SCG-HSCG Adder

Usthulamuri Penchalaiah[1] and V.G. Siva Kumar[2]

ABSTRACT

In advancement technology and applications like processing sound signals, processing image signals, software define radio signals and so on. In processing of digital systems in all type of devices may be wide range of contraptions and clamors. The result of design the all the athematic devices and performing the addition and multiplication operations need larger space in very large scale implementation. Here, the latest proposed methodology progressed in the design of Finite impulse response filter by redesign the multipliers and adders to introduce new unsigned and signed truncated multiplier and new SCG-HSCG adders. Also different III-V[th] semiconductor materials can be used to get high speed filtering operation. In this way to decreases the partials part reduction of design truncated multiplier have more number of adders. The new design handles to replace the SCG-HSCG Adders. This design decreases the logic gates in operations of athematic, additions and multiplications. It also gives the more efficient in the sign and unsigned designs of different digital signal processing applications. This proposed new finite impulse response filter has designed with help of VHDL and Xilinx FPGA-S6LX9. The efficiency of proposed filter with truncated multiplier and SCG-HSCG adder, is compare with the existing all CSLA – Carry Select adder in terms of area, delay, and power. The novel FIR filter design reduces power distribution to the environment while increasing hardware device performance. Furthermore, these kinds of designs take up less space, which means they use less energy and last longer.

Keywords: *Half Sum Carry Generation, Sum Carry Selection, FPGA, III-V[th] Semiconductor Materials.*

---

[1] *Presenting & corresponding author.*
Research scholar Department of Electronics and Communication Engineering, Sathyabama Institute of Science and Technology, Chennai 600119, India. ***E-mail:*** penchal.upr@gmail.com
[2] Department of Electronics and Communication Engineering, Vidya Jyothi Institute of Technology, Hyderabad 5000759, India.

F-EIR Conference 2021 on
Environment Concerns and its Remediation (ECR21)
Chandigarh, India, October 18–22, 2021

Compound Wet Hazards Intensify the Vulnerability of Indian Himalayan Region: A District-Level Assessment to Suggest Suitable Adaptation Models

Akshay Singhal[1], Muhammed Jaseem[2], M. Kavya[2], Sanjeev Kumar Jha[3] and Vinee Srivastava[2]

ABSTRACT

Compound wet hazards usually have more potential to create severe widespread impacts than any single hazard alone. Generally, hazard assessment methods consider only one hazard at a time. Such assessments may lead to underestimation of actual impacts, especially when the drivers of the hazards interact in a common space and time. The Himalayan region, due to its complex geology, topography, uncertain weather patterns and increased anthropogenic activities is more prone to the devastating impacts of compound hazards. In this study, we conduct a vulnerability assessment of the Indian Himalayan Region (IHR) due to compound wet hazards (CWH) induced by a common precursor, i.e., extreme rainfall. We consider four types of hazards induced by extreme rainfall such as floods, cloudbursts, avalanches and landslides. The study area consists of 87 districts spanning across 11 states of the IHR. The vulnerability of each district is identified and classified on the basis of its susceptibility to the number of hazards and also on possible combinations of hazards. To assess the severity of the hazards, an integrated approach is adopted to inform the model with data of various indicators belonging to weather, topography, vegetation and local socio-economic conditions. Each indicator is allotted the dimensions of exposure, sensitivity and adaptive capacity based on its nature of function. The indicators are then assigned appropriate weights and subsequently aggregated to form compound hazard vulnerability index (CHVI). Outcomes of the study are used to classify districts with single and compound hazards and associated cause(s) are identified. Moreover, a scheme of compound hazard adaptation model (CHAM) is proposed for the districts prone to multiple hazards based on their vulnerability status. Such a vulnerability assessment which identifies the Himalayan districts prone to multiple natural hazards,

[1] *Presenting author*
 Indian Institute of Science Education and Research Bhopal, India.
[2] Indian Institute of Science Education and Research Bhopal, India.
[3] *Corresponding author*
 Indian Institute of Science Education and Research Bhopal, India.
 E-mail: sanjeevj@iiserb.ac.in

classifies their hazard status based on impacts, evaluates the possible causes and suggest suitable adaptation models, can assist national and state disaster management agencies in reducing potential risks. Moreover, efficient mitigation measures can also be implemented based on suggested adaptation models.

Keywords: Himalayas, Vulnerability, Landslide, Flood, Avalanche, Cloudburst.

Graphical Abstract

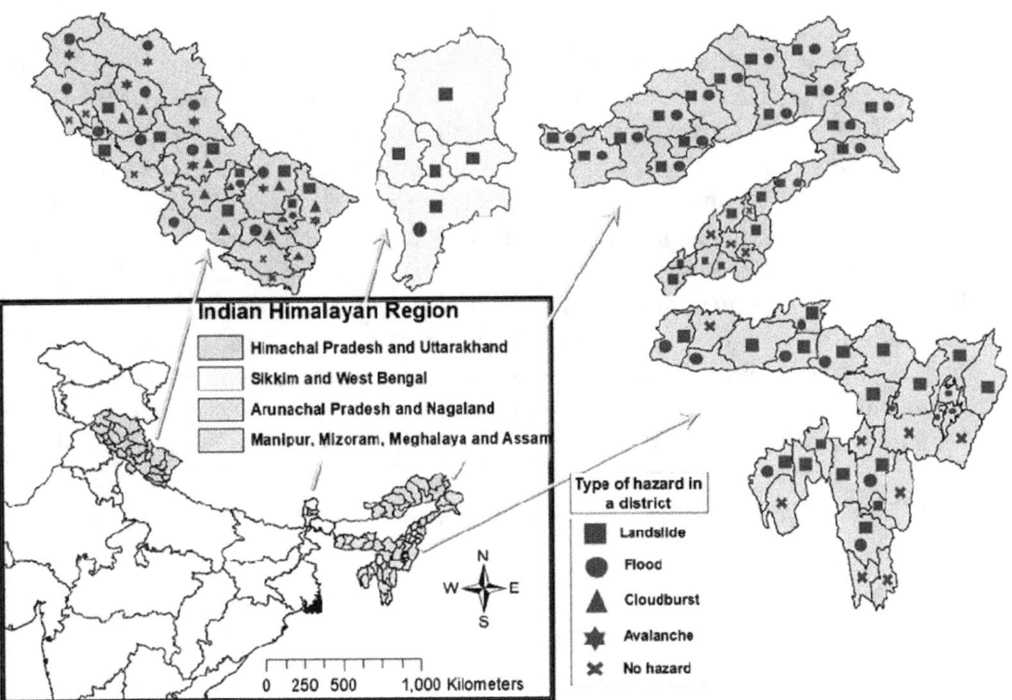

F-EIR Conference 2021 on
Environment Concerns and its Remediation (ECR21)
Chandigarh, India, October 18–22, 2021

Effect of Cell Concentration for Biological Treatment of RCA

Himanshu Sharma[1] and Sanjay Sharma[2]

ABSTRACT

The growing urge for the utilization of construction and demolition waste recycled concrete aggregate (RCA) in the concrete demands for substantial improvement in the quality of RCA. Various treatment technologies have been applied by researchers in past decade, of which, biological treatment of RCA by utilizing calcium secreting micro-organisms has been found more effective. This study focuses on the effect of cell concentration of bacillus sphaericus on the various parameters of RCA. Four culture mediums with cell concentration of 1×10^2 to 1×10^8 cfu/ml were taken for the treatment of RCA. After the treatment various properties of aggregate like water absorption, weight increase, density, strength, specific gravity, crushing value and impact value have been studied. After the analysis, it was observed that improvement in the properties of RCA increased with increase in cell concentration from 1×10^2 to 1×10^6 cfu/ml, but as the cell concentration increased to 1×10^8 the improvement in the properties of the RCA reduced.

Keywords: Biological Treatment, Bacillus Sphaericus, Optimum Cell Concentration.

[1] *Presenting & corresponding author.*
Research Scholar, Department of Civil Engineering, I.K. Gujral Punjab Technical University, Kapurthala 144603, India. *E-mail:* Himanshu1027@gmail.com
[2] Professor, Department of Civil Engineering, National Institute of Technical Teachers Training and Research, Chandigarh 160019, India. *E-mail:* sanjaysharmachd@yahoo.com

F-EIR Conference 2021 on
Environment Concerns and its Remediation (ECR21)
Chandigarh, India, October 18–22, 2021

Influence of Various Treatment Technologies on the Various Parameters of RCA

Himanshu Sharma[1] and Sanjay Sharma[2]

ABSTRACT

Rapid growth in economy has given a substantial rise to the demand of construction resulting in the increased construction and demolition waste (CDA). Utilization of recycled aggregate in concrete may help in reducing it, but the properties of recycled concrete aggregate (RCA) are very much inferior in comparison to the natural aggregate (NA) and thus cannot be utilized directly into the concrete. To solve this problem various researchers have developed different kind of techniques to improve the quality of RCA. In this study a comparative analysis of various treatment technologies have been performed to improve the quality of RCA. This study focuses on the parameters like, water absorption, density, adhered mortar, strength, durability and microstructure of RCA. After studying various treatment technologies for RCA, it can be summarized that the techniques like carbonation, microbial treatment and utilization of nano materials as best suited for improving the new and old interfacial transition zones of RCA, which in turn can improve many properties like, water absorption, density, crushing strength, impact strength, micro cracks.

Keywords: Recycled Aggregate Concrete, Construction and Demolition Waste, Carbonation, Microbial Treatment, Interfacial Transition Zone.

[1] *Presenting & corresponding author.*
Research Scholar, Department of Civil Engineering, I.K. Gujral Punjab Technical University, Kapurthala 144603, India. *E-mail:* Himanshu1027@gmail.com
[2] Professor, Department of Civil Engineering, National Institute of Technical Teachers Training and Research, Chandigarh 160019, India. *E-mail:* sanjaysharmachd@yahoo.com

F-EIR Conference 2021 on
Environment Concerns and its Remediation (ECR21)
Chandigarh, India, October 18–22, 2021

Sustainable Construction Practices: A Perspective View of Indian Construction Industry Professionals

Mag Raj Gehlot[1] and Sandeep Shrivastava[2]

ABSTRACT

The building and construction industry is the crucial responsible element for destroying the environment by excessive resource consumption and producing waste. The extensive development has knocked on the door of experts and industry professionals to compete with the consequences by adopting sustainable practices in building construction projects. As a result, this study aimed to address the awareness and factors that resist the adoption of sustainable construction practices in India. Through a questionnaire survey, practicing professionals filled the data from pan India construction projects. The analysis conducted on the data received from 62 projects, among 120 selected, where 56% respondents were site engineer and 16% respondent were Project managers. The selected projects had 12% residential whereas remaining were non-residential and infrastructural projects. The results revealed three significant findings that influence the adoption of sustainable construction practices: first, the lack of skilled human resources availability. Second, the reluctant client. Third, limited and lack of knowledge of sustainable technology and operation to the professionals as these three major factors had the mean item score (MIS) as 3.28, 3.14 and 3.12 respectively. Hence, the study contributes towards the sustainability thinking of the various stakeholders of the Indian construction industry. The finding suggested that simplified strategies and participative efforts need to be conveyed to all the stakeholders of the particular project and industry practitioner to boost the process of creating sustainable environments for the building construction industry, which can essentially help in creating a sustainable future.

Keywords: Sustainable Construction Practices, Construction Industry Professionals, Sustainable Constructions, Sustainable Materials.

[1]*Corresponding & Presenting author.*
Assistant Professor, Department of Civil Engineering, UCET, Bikaner, India.
E-mail: m.rajgehlot@gmail.com
[2]Assistant Professor, Department of Civil Engineering, MNIT Jaipur, Jaipur, India.
E-mail: san.civil@gmail.com

F-EIR Conference 2021 on
Environment Concerns and its Remediation (ECR21)
Chandigarh, India, October 18–22, 2021

Assessment and Suitability Analysis of Water Quality of River Ganga in Patna, Bihar

Reena Singh[1] and Saurabh Kumar[2]

ABSTRACT

The river Ganga is the largest river in India. It is the number one resource for ingesting water for humans in villages, towns, and cities. Its water is used for many purposes like drinking, agriculture, cattle bathing, etc. People's regular use of Ganga water for all different works increases the possibility of human fitness dangers. This paper deals with the identity of the water quality status of selected fourteen ghats in Patna. For this, two types of Water Quality Index (WQI) have been used namely the Weighted Arithmetic Mean method and the Oregon Water Quality Index method. For calculation of WQI, the physicochemical parameters namely Temperature, Electrical conductivity (EC), pH, Total Dissolved Solids (TDS), Alkalinity, Hardness, Chloride, Turbidity, Dissolved Oxygen (DO), and Biological Oxygen Demand (BOD) are determined. The samples were collected from Feb 2019 to February 2020 just before complete lockdown. It was found in the study that the WQI of all the ghats was categorized as unsafe for drinking purposes and suggested that water should be supplied after proper treatment.

Keywords: *River Ganga, Water Quality Index, Water Quality Parameters, Unsafe for Drinking.*

[1] *Presenting & corresponding author.*
Department of Civil Engineering, National Institute of Technology Patna, India.
E-mail: reena@nitp.ac.in
[2] Department of Civil Engineering, National Institute of Technology Patna, India.

Author Index